Advances in Geographical and Environmental Sciences

Advances in Geographical and Environmental Sciences synthesizes series diagnostigation and prognostication of earth environment, incorporating challenging interactive areas within ecological envelope of geosphere, biosphere, hydrosphere, atmosphere and cryosphere. It deals with land use land cover change (LUCC), urbanization, energy flux, land-ocean fluxes, climate, food security, ecohydrology, biodiversity, natural hazards and disasters, human health and their mutual interaction and feedback mechanism in order to contribute towards sustainable future. The geosciences methods range from traditional field techniques and conventional data collection, use of remote sensing and geographical information system, computer aided technique to advance geostatistical and dynamic modeling.

The series integrate past, present and future of geospheric attributes incorporating biophysical and human dimensions in spatio-temporal perspectives. The geosciences, encompassing land-ocean-atmosphere interaction is considered as a vital component in the context of environmental issues, especially in observation and prediction of air and water pollution, global warming and urban heat islands. It is important to communicate the advances in geosciences to increase resilience of society through capacity building for mitigating the impact of natural hazards and disasters. Sustainability of human society depends strongly on the earth environment, and thus the development of geosciences is critical for a better understanding of our living environment, and its sustainable development.

Geoscience also has the responsibility to not confine itself to addressing current problems but it is also developing a framework to address future issues. In order to build a 'Future Earth Model' for understanding and predicting the functioning of the whole climatic system, collaboration of experts in the traditional earth disciplines as well as in ecology, information technology, instrumentation and complex system is essential, through initiatives from human geoscientists. Thus human geoscience is emerging as key policy science for contributing towards sustainability/survivality science together with future earth initiative.

Advances in Geographical and Environmental Sciences series publishes books that contain novel approaches in tackling issues of human geoscience in its broadest sense — books in the series should focus on true progress in a particular area or region. The series includes monographs and edited volumes without any limitations in the page numbers.

Anupama Dubey · Subhash Anand ·
Basavaraj Bagade
Editors

Promoting Sustainability Through Water Management and Climate Change Adaptation

Springer

Editors
Anupama Dubey
Institute of Public Enterprise
Osmania University Campus
Hyderabad, Telangana, India

Subhash Anand
Department of Geography
University of Delhi
New Delhi, Delhi, India

Basavaraj Bagade
Department of Studies and Research
in Geography
Rani Channamma University
Belagavi, Karnataka, India

ISSN 2198-3542 ISSN 2198-3550 (electronic)
Advances in Geographical and Environmental Sciences
ISBN 978-981-99-5478-0 ISBN 978-981-99-5479-7 (eBook)
https://doi.org/10.1007/978-981-99-5479-7

This Springer imprint is published by the registered company Springer Nature Singapore Pte Ltd.
The registered company address is: 152 Beach Road, #21-01/04 Gateway East, Singapore 189721, Singapore

Preface

Water is the primary channel through which people experience the effects of climate change. In many regions, water availability is becoming less predictable, and flooding is more frequent, posing a threat to destroy water points and sanitation facilities as well as pollute water sources. Societies have always been obliged to handle the natural unpredictability in their water sources. In addition to modifying and aggravating extreme weather patterns and increasing long-term uncertainty in water supply quantity and quality, climate change threatens to exacerbate this unpredictability. In addition to demographic, economic, environmental, social, and technical challenges, climate change is a major element impacting the management of water supply. Water drives the environment, social, and economic development pillars of sustainable development. According to the United Nations Sustainable Development Goals (SDGs), the development of these pillars is contingent upon the availability and management of water resources to meet the demand for water. In a climate that is continuously changing, the growing significance of water resources is vital for attaining sustainable development.

This Volume has been dedicated to explain different methods adopted and various issues discussed in the area of water resource management and part explains the chapters dedicated on different issues on climatic adaptation.Chapters such as, Livelihood change and sustainability potential in a Sri Lankan mountain village, Geostatistical Modeling and Mapping of As Occurrence and Vulnerability, Ensuring potable water supply to rural areas, Study of Spatio-Temporal Variation in Rainfall at Suketi River Basin by using Rainfall Anomaly Index (RAI), Suitability of Groundwater for Irrigation, Reinforcement of Drinking Water in Fluoride Affected Areas of Nalgonda District Through Improvised Rainwater Harvesting System and Estimation of surface runoff usingNRCS CN method and Geospatial Techniques for sub-basins prioritization of conservation planning of Ghera Sinhagad Land System, Western Maharashtra depicts the diversity of methodology adopted by the authors to study the mentioned issues in the domain of water resource management. This book is unique in nature as a chapters also elobartes the water conflicts and regulation of water resources through a comparative assessment of two South-Indian states. Role of gender in Water Resource Management has fetched attention of international Policy makers and hence plays a crucial role in SDG Goals 2030. This book covers the role of gender in water resource management by taking case study from the North Eastern State of India. Second part discusses various climatic and Environmental issues such as, Assessing Human-wildlife conflicts in Tiger corridor habitat, Environmental Ethics": In the Context of the COVID-19, BuildingSustainable Livelihoods Through Everyday "Green, Urbanism" Practices—A case study of GTB Nagar Neighborhood Delhi, India., Understanding the Impact of Climate Change and Policy Development in India during Post-NAPCC Era, Valuing Benefits of Urban Green Spaces for Mitigation of Climate Change Impacts and Promoting Urban Resilience.

Thus, in nutshell this volume is an attempt to understand various relevant issues pertaining to the theme and provoke the necessary attention to resolve the problems by adopting the innovative and scientific techniques.

Hyderabad, India — Anupama Dubey
New Delhi, India — Subhash Anand
Belagavi, India — Basavaraj Bagade

Acknowledgements

Each Chapter contributor of this book has made substantial contributions to the fields of Sustainable Development, Water Management, and Climate Change. Without their efforts, this book could not have been drafted to this version, thus we gratefully thank all the professors, researchers, scientists, water and climate change specialists, managers, and administrators who have contributed in this connection. We appreciate the language editor from the Institute of Public Enterprise in India for his insightful comments.

We appreciate the assistance and support from the Springer team.

December, 2022

Anupama Dubey
Subhash Anand
Basvaraj Bagade

Contents

Part I
Sustainability Through Water Resource Management

Chapter 1
Geostatistical Modeling and Mapping of Arsenic Occurrence and Vulnerability—A Case Study on Bihar, India

Sana Dhamija and Himanshu Joshi

Abstract Determination of the vulnerability of groundwater to arsenic contamination in general and mapping the vulnerable zones is essential for the preservation of groundwater quality. CGWB has explored and marked the occurrence of arsenic mainly along the Ganga–Brahmaputra-Meghna (GBM) river basin. The major aim of the current study is to conduct geostatistical modeling of arsenic occurrence in the groundwater employing the available secondary data based on the subsurface and hydrogeochemical parameters for the districts of Bihar. Further, the arsenic vulnerability index has been developed employing the Multi-Criteria Decision Making (MCDM) method after integrating various thematic layers in the GIS platform and generating a vulnerability map. This has helped to identify the arsenic-vulnerable zones. The north-western and the central parts, including Patna, Bhojpur, Munger, Bhagalpur, and Vaishali districts, exhibit high arsenic risk zones due to shallow depth of water and low elevation associated with high silicate and iron contents. It is found that high arsenic risk is associated with the districts lying along the Ganga River. The output was validated against the existing arsenic locations for 2015 (CGWB) and, 2019 (primary data of three districts named Saran, Samastipur, and Vaishali), and by the Receiver Operating Characteristics (ROC) metric.

Keywords Arsenic contamination · Arsenic vulnerability · Arsenic Vulnerability Index (AVI) · Analytical Hierarchy Process (AHP)

S. Dhamija (✉) · H. Joshi
Department of Hydrology, Indian Institute of Technology Roorkee, Roorkee, Uttarakhand 247667, India
e-mail: sdhamija@hy.iitr.ac.in

A. Dubey et al. (eds.), *Promoting Sustainability Through Water Management and Climate Change Adaptation*, Advances in Geographical and Environmental Sciences,
https://doi.org/10.1007/978-981-99-5479-7_1

1.1 Introduction

Deterioration in groundwater quality has increased in India following its over-exploitation due to greater dependency on groundwater in all sectors. In this regard, arsenic toxicity due to groundwater has been acknowledged as a major health concern, even at its low concentrations. Arsenic is believed to spread in the food chain through irrigation in addition to being consumed as drinking water and used for other household tasks. Arsenic contamination has spread throughout a significant portion of the flood plains in the Ganges and Brahmaputra basins (Dhamija and Joshi 2022). Several hypotheses have suggested the geogenic (weathered igneous and sedimentary rocks), anthropogenic (mining, smelting, refining, and industrial processes), and biogenic (bioaccumulation and biotransformation) sources for arsenic release in groundwater (Jang et al. 2016; Kumar 2015; Matschullat 2000; Nordstrom 2002). Arsenic, incidentally, is tasteless and odorless and has remained unrecognized for years. It is imperative that an assessment of groundwater vulnerability to arsenic is an urgent need of the time.

Arsenic occurrence is displayed in organic (MMA(III), MMA(V), DMA(III), DMA(V)), and inorganic (As^{3-}, As^{0}, As^{3+}, and As^{5+}) forms in the earth's crust. Among all iAs forms, i.e., arsenite (As^{3+}) and arsenate (As^{5+}) are mobile, which undergo microbial transformations, and are extremely toxic (Chatterjee et al. 2017; Wang and Mulligan 2006). Drinking water acts as a major route for arsenic exposure to humans. BIS (2012) declared 0.1 mg/l as the acceptable limit for As in drinking water with a maximum limit of up to 0.5 mg/l, which was later amended to 0.1 mg/l with no relaxation (WHO, 2022). CGWB (2015) has reported that fifteen districts of Bihar are contaminated with groundwater arsenic. This chapter deals with the twelve districts of Bihar lying in the vicinity of the river Ganga.

The principal requirement for arsenic vulnerability assessment is the evaluation of all the possible factors (or covariates) responsible for arsenic mobilization. It is a complex task as multiple covariates are selected based on hydro-geological and hydro-chemical attributes. The standards for the selection of these are neither specific nor fixed as each one has a unique significance. Further, researchers employ many techniques to determine the priority weights for these before aggregating them. In the present study, attempts have been made to categorize the arsenic vulnerable risk zones using the GIS-based Analytical Hierarchy Process (AHP) approach. Saaty (1977) developed AHP as one of the traditional Multi-Criteria Decision Making (MCDM) methods which evaluates the weights based on a pair-wise comparison matrix (Saaty and Vargas 2006). MCDM has the power to show real results because it can handle huge datasets by allocating weights and rating to them (Agarwal and Garg 2015).

1.2 Study Area

In Bihar, arsenic was first reported in groundwater in 2002 (Chakroborti et al. 2011). The state covers a range of latitude from 24°20′10″ N ~ 27°31′15″ N and longitude from 82°19′50″ E ~ 88°17′40″ E (GOI 2015). M/o WR and GR (2014) investigated a wide range of As contamination in 57 blocks of 15 districts. The present study covers the twelve arsenic-affected districts of the state near the Ganges, covering 26,558 sq. km of area (Fig. 1.1). The average elevation of the state is around 53 m above sea level. The state is positioned between West Bengal (humid) and Uttar Pradesh (semi-humid). Hence, it has a transitional climate with 100–150 cm annual rainfall. Bihar has fertile agricultural land with various cereals and crops grown. The current area has a total population of 3,50,96,572 persons (Census 2011).

The reasons behind selecting the study area for research are: (i) The selected districts lie in the vicinity of the main Ganga river (Fig. 1.1). (ii) Groundwater of the selected districts has high As contamination. (iii) Bihar is a highly productive state with a major occupation being agriculture, with irrigation usually practiced with As-contaminated water. (iv) The area is densely populated, which propels the need for major groundwater extraction along with the build-up of organics.

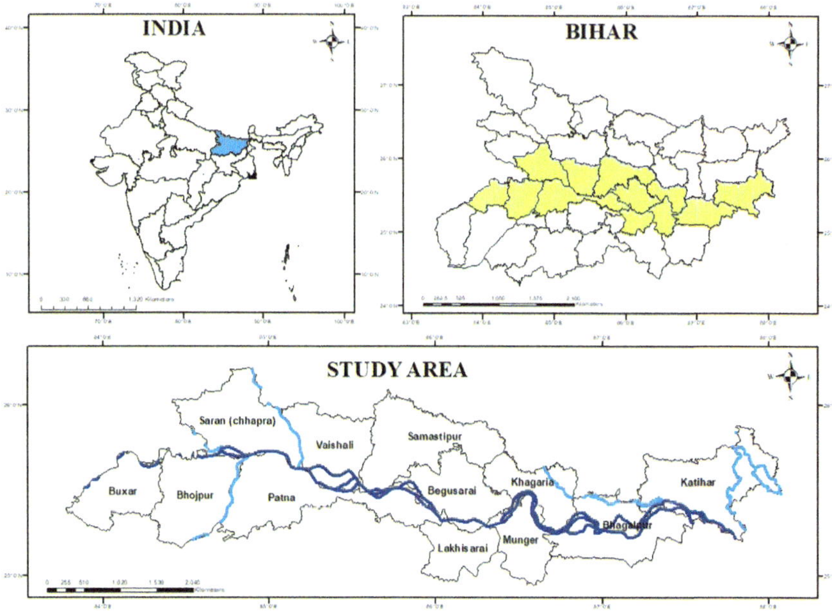

Fig. 1.1 Study area—Bihar, India

1.3 Methodology

The methodology is described in the following flowchart (Fig. 1.2).

1.3.1 Data Procurement

Relevant secondary data and available information (i.e., depth to groundwater, geomorphology, DEM images, and groundwater quality parameters) were collected from the official documents and reports prepared by the government/ non-government organizations; and reported on relevant websites (Table 1.1). Procured datasets were processed and analyzed using ArcGIS v. 10.6.1. software.

1.3.2 Analysis and Modeling

Analysis and modeling were performed by applying the Analytical Hierarchy Process (AHP) approach of the Multi-Criteria Decision Making (MCDM) method. MCDM is used for conflicting several decision-making criteria (Keshavarz-Ghorabaee et al. 2018). AHP is a systematic way of decision-making based on preferences given to the alternatives (Levary 2008). It uses hierarchical structures to represent a problem and then develop priorities for the alternatives based on the user's judgment (Saaty 2008). The following steps were performed for AHP analysis (Lee et al. 2008): (1)

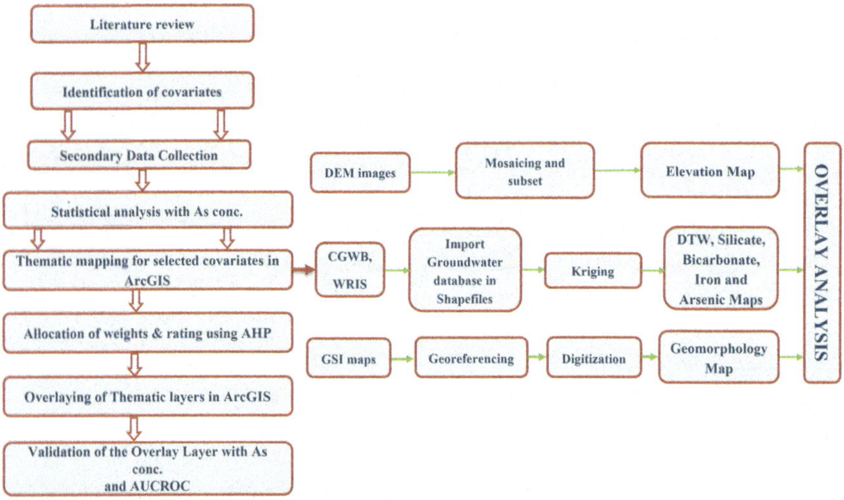

Fig. 1.2 Schematic flowchart of methodology

Table 1.1 Selected covariates and the data sources

S.No	Covariate	Data source	Format	Year
1	Geomorphology	Geological Survey of India (GSI) https://www.gsi.gov.in/webcenter/portal	Map	2007
2	Depth to water level	Water Resource Information System (WRIS) https://indiawris.gov.in/wris/#/DataDownload	Tabular	2015
3	Elevation	United States Geological Survey (USGS) https://earthexplorer.usgs.gov/	SRTM DEM Imagery	2015
4	Groundwater quality parameters	Central Ground Water Board (CGWB) http://cgwb.gov.in/wqreports.html	Tabular	2015

The unstructured problem was outlined; (2) Related covariates were selected; (3) The hierarchical framework based on the selected covariates was created; (3) The weights were determined using the pair-wise comparison matrix between the input factors via Saaty's Ratio Scale (Table 1.2) employing eigenvalue technique based on the literature surveys (Triantaphyllou and Mann 1990); (4) The consistency ratio (CR) was evaluated, which must be less than 0.1 (Saaty 2008); (5) Overall priority ranking was allocated, and the best alternative was selected.

Table 1.2 Saaty's ratio scale

Intensity of importance	Definition	Explanation
1	Equal importance	Two activities contribute equally to the objective
3	Weak importance of one over another	Experience and judgment slightly favor one activity over another
5	Essential or strong importance	Experience and judgment strongly favor one activity over another
7	Demonstrated importance	An activity is strongly favored and its dominance is demonstrated in practice
9	Absolute importance	The evidence favoring one activity over the another is of the highest possible order of affirmation
2,4,6,8	Intermediate values	When compromise is needed

The Consistency Ratio (CR) check was achieved by the following Eq. 1.1 (Dhamija and Joshi 2022):

$$\mathrm{CR} = \frac{\mathrm{CI}}{\mathrm{RCI}} \tag{1.1}$$

where CR = Consistency Ratio;

CI = Consistency Index;
RCI = Random Consistency Index.

CI here is obtained by Eq. 1.2 (Dhamija and Joshi 2022):

$$\mathrm{CI} = \frac{(\lambda\mathrm{max} - n)}{(n - 1)} \tag{1.2}$$

where CI = Consistency Index;

λmax = Maximum Eigenvalue of the matrix;
n = number of groundwater-affecting covariates.

A dimensionless Arsenic Vulnerability Index (AVI) was used to integrate all characteristics into a single map, and final arsenic vulnerability maps were generated. AVI was evaluated using the total of the products of ratings and weights assigned to each covariate (Eq. 1.3) (Dhamija and Joshi 2022).

$$\mathrm{AVI} = \mathrm{G_r G_w} + \mathrm{D_r D_w} + \mathrm{E_r E_w} + \mathrm{Si_r Si_w} + \mathrm{B_r B_w} + \mathrm{Fe_r Fe_w} \tag{1.3}$$

where G = Geomorphology

D = Depth.
E = Elevation.
Si = Silicate.
B = Bicarbonate.
Fe = Iron.
r = rating.
w = weightage.

The quantile and natural jerk classification methods in ArcGIS are used to classify the high and low vulnerability areas on the arsenic vulnerability map (Fig. 1.3a–f).

1.3.3 Data Validation

The accuracy of the output was determined by validating the generated vulnerability map with the groundwater arsenic concentration database of three districts for the years 2015 (secondary data) and 2019 (primary data). Furthermore, the Receiver

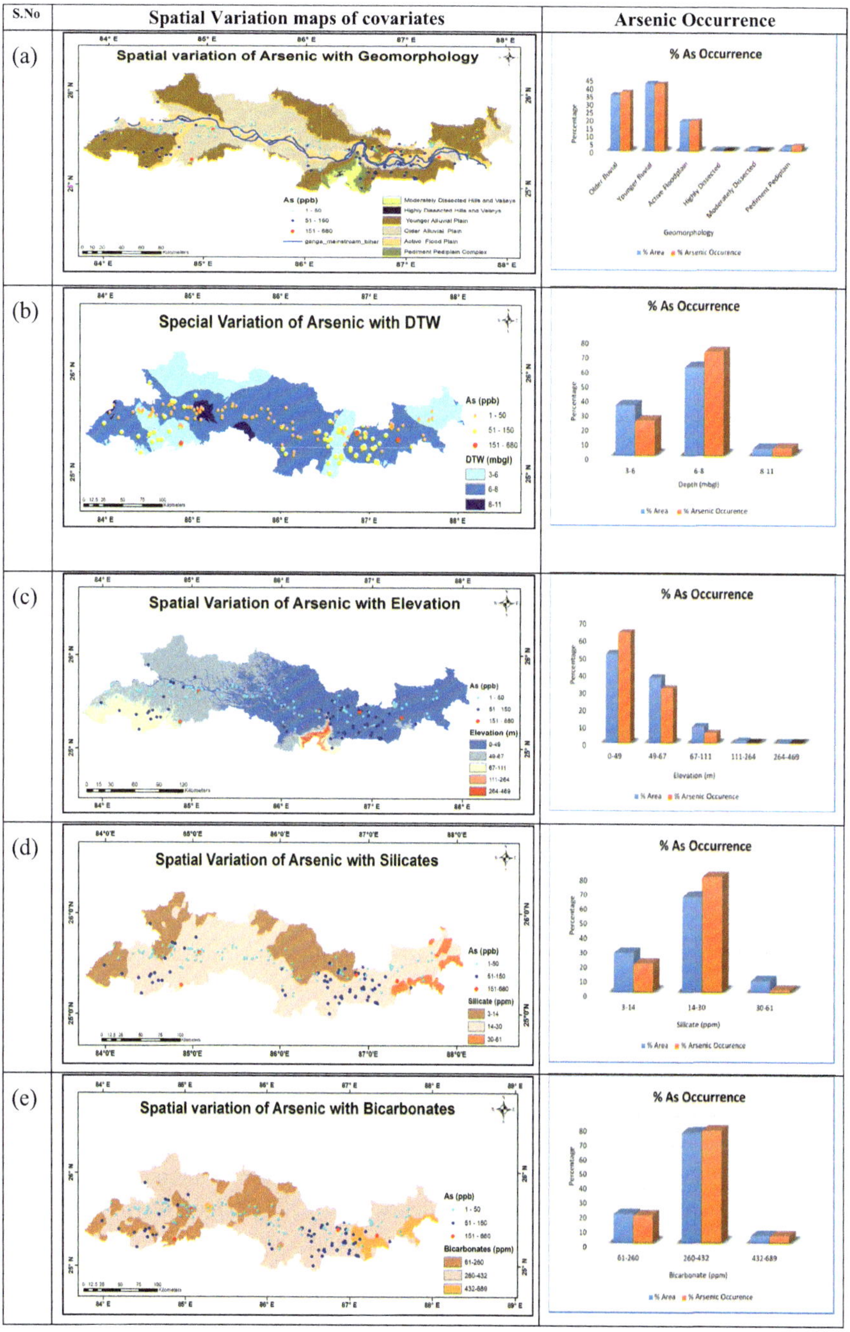

Fig. 1.3 Thematic layers and Arsenic Occurrence percentages w.r.t. (a) Geomorphology; (b) DTW; (c) Elevation; (d) Silicates; (e) Bicarbonates; (f) Iron concentration of the study area

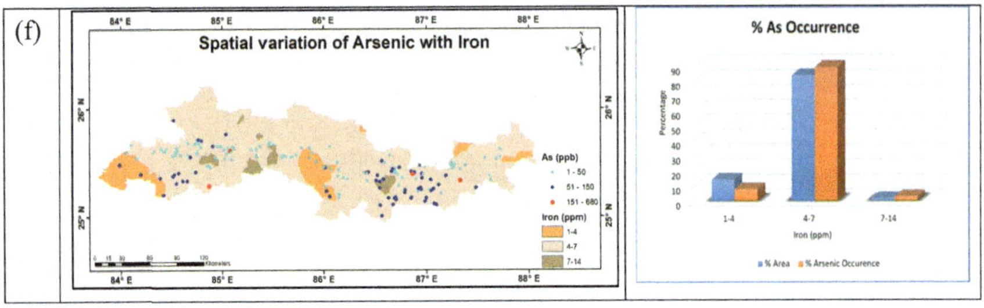

Fig. 1.3 (continued)

Operating Characteristics (ROC) curve analysis was computed using an arsenic training dataset and the arsenic vulnerability map produced.

1.4 Results and Discussion

Various hypotheses govern the geogenic occurrence of arsenic in the groundwater because of the abundance of arsenic-bearing mineral reservoirs. The Himalayas are considered the principal source of arsenic in the groundwater of the Ganga-Meghna-Brahmaputra plains (Chakroborti et al. 2011). In the present study, selected covariates were considered for preparing the arsenic vulnerability maps using AHP and AVI, and were later classified into low, moderate, and high vulnerability.

1.4.1 Assessment of as Occurrence

As occurrence was evaluated by geostatistical modeling, presented graphically (Fig. 1.3a–f). The spatial relationship was examined by overlaying the thematic layer of groundwater As concentration with selected covariates (geomorphology, water table, elevation, and groundwater quality parameters). The relationship of arsenic with its covariates was established statistically and represented by bar graphs of the percentage of arsenic occurrence, which were created concerning the area covered by each covariate (Fig. 1.3a–f).

The geological evidence for groundwater contamination with arsenic is symbolized by geomorphology. Alluvial plains deposited by the Ganges and its tributaries, typically cover the study area. It was noted that arsenic contamination was anticipated in the younger alluvium plains of the state because younger alluvium covers 42% of the land. The quaternary sediments of the alluvial stratigraphy have a significant association with the distribution pattern of arsenic in the districts, which is abundant in minerals like quartz, feldspars, illite, and kaolinite along with organic matter from fine-grained rocks (Bhattacharya et al. 1997; Das et al. 1995; Nickson et al. 1998).

The younger alluvium has a high proportion of arsenic since it is reductive in nature and is rich in organic materials, as shown by the bar graph between the area % of the AHP method and the occurrence of arsenic (Fig. 1.3a).

It was found that the potential aquifers of these districts are shallow (<10 m) (Fig. 1.3b) and have a significant content of arsenic when the depth to water level was taken into account. According to Saha et al. (2009), arsenic concentration declines with depth, and the current study predicted the same. The reason may be the oxidation of arsenic-rich iron sulfide minerals. Secondly, when an aquifer is heavily pumped for agricultural purposes, the surrounding clay and water become less compact as the land descends. This could cause groundwater that is contaminated with arsenic to accumulate in deep aquifers (Smith et al. 2018).

The majority of the study area under consideration is low-lying with a high concentration of arsenic facilitated by the build-up of finer sediments in low-elevated areas. Low-sloped areas hold onto water for a long time, which increases infiltration, water recharge, and the risk of groundwater contamination and vice-versa (Lynch et al. 1994). A high proportion of arsenic is also present in relatively flat locations, as shown by the bar graph, which suggests that the accumulation of finer sediments in low-lying places causes a significant arsenic infiltration into deep aquifers (Fig. 1.3c). As a result, this infiltration identifies the anthropogenic source of arsenic in groundwater (least contributed) (Table 1.3).

Arsenic displayed a high correlation with the groundwater quality parameters like silicate, bicarbonates, and iron, as per the data obtained from CGWB. As per the data, silicate concentration ranges from 3–52 ppm. Elevated concentrations of silicates (<15 ppm) and As (>50 ppb) were observed in Bhagalpur, Bhojpur, Katihar, Patna, Munger, Samastipur, and Vaishali districts. Also, these districts lie in the region of newer alluvium plains and active flood plains, which is responsible for creating reducing conditions in the aquifer. Silicates being a major fraction of sediments, upon leaching at high pH and low redox potential, release arsenic during silicate mineral dissolution (Alam et al. 2014). Bar graph also indicated the elevated As with high silicate content, which accelerates more silicate mineral dissolution and releases arsenic (Fig. 1.3d).

As per the data, bicarbonate concentration ranges from 61 to 687 ppm. Higher bicarbonate (>20 ppm) was reported in all districts except a few parts of Begusarai,

Table 1.3 AHP pairwise comparison for Bihar districts

Covariates	D	E	G	Si	B	Fe	Weightage (%)
Depth (D)	0.14	0.17	0.12	0.11	0.22	0.14	15.00
Elevation (E)	0.07	0.08	0.06	0.07	0.06	0.14	8.00
Geomorphology (G)	0.14	0.17	0.12	0.11	0.06	0.14	12.00
Silicate (Si)	0.29	0.25	0.24	0.21	0.33	0.14	24.00
Bicarbonate (B)	0.07	0.17	0.24	0.07	0.11	0.14	13.00
Iron (Fe)	0.29	0.17	0.24	0.43	0.22	0.29	27.00

CR = 0.049 < 0.1

Buxar, Patna, and Samastipur (Fig. 1.3e). Considering As concentration, quite similar pattern with few more exceptions like parts of Katihar, Khagaria, and Saran were observed. High As in these areas is due to bicarbonate ions, which can extract arsenic from the sediment samples in oxic and anoxic conditions. Bar graph also represents that elevated As is observed in association with high bicarbonates (>260 ppm), suggesting that bicarbonate ions have an affinity to form complexes on the surface sites of iron hydroxide and substitute arsenic from the surface of minerals and sediments, resulting in the release of arsenic to groundwater (Saha et al. 2009).

As per the data, iron concentration ranges from 1 to 13 ppm. Higher iron was reported uniformly in the state except in some parts of the districts such as Bhojpur, Katihar, Lakhisarai, and Samastipur (a few) (<4 mg/l) (Fig. 1.3f). In the dry season, weathering of clay sediments immobilizes (less mobile) iron for long durations, resulting in the release of mobile arsenic due to its strong affinity with pyrites. Similarly, Fe–Mn oxyhydroxides are a minor component of sediments, but are larger reservoirs of arsenic, undergo reductive dissolution, and release As in the groundwater (Alam et al. 2014). Literature also suggests that organic carbon encourages microbial respiration, which further activates the reductive dissolution of As and Fe in the solid phase (Neidhardt et al. 2014).

1.4.2 Assessment of Arsenic Vulnerability

To evaluate arsenic vulnerability, weights and ratings were assigned to all covariates and classes of each covariate based on a pair-wise comparison of the AHP (Table 1.3). For the current study, CR was 0.049 (must be less than 0.1, hence accepted). Further, reclassification was performed for each covariate thematic layer which was combined in a single map using AVI. The final arsenic vulnerability map was obtained by the overlay analysis tool of ArcGIS, which was represented by quantile (Fig. 1.4a) and natural classification methods (Fig. 1.4b). Considering the permissible limit, As vulnerability was categorized into three classes, low (<50 ppb), moderate (50–100 ppb), and high (>150 ppb). Both the classification methods have shown similar results with minimal differences. The resultant patchy patterns justify the ground situation to a larger extent. The index value ranges from 113 to 295 for the study area. The highest index value has been displayed around and along the Ganges, with the northern regions of Buxar, Bhojpur, Bhagalpur, and Munger districts; the southern part of Saran, Vaishali, Lakhisarai, Begusarai, and Katihar are at high risks of approx. 27% (Fig. 1.4a, b, c). Areas with the shallow water level lying in the alluvial channel zone show low and moderate arsenic vulnerability.

Bihar is primarily covered with younger, older, and flood alluvial plains, with 42% of arsenic contamination in younger alluvial plains due to the accumulation of arsenic by organic-rich silt, clay, and sand (Matisoff et al. 1982) (Fig. 1.3a). Oxidation of arsenic-rich iron sulfide minerals releases arsenic which is observed in 69% of the potential aquifers of shallow depths (levels < 10 m) (Fig. 1.3b). Low surface elevation accumulates fine sediments with arsenic-adsorbing minerals

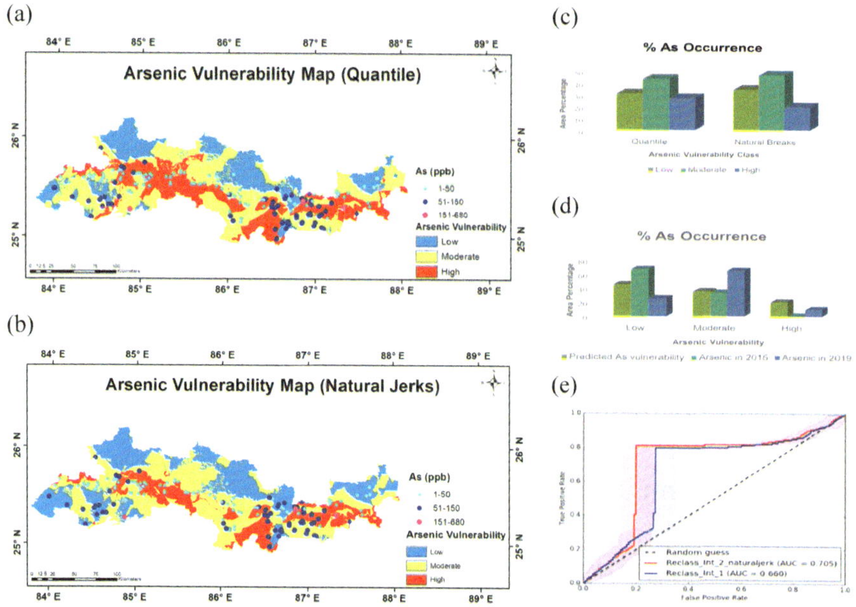

Fig. 1.4 Arsenic Vulnerability Map for Bihar by (a) Quantile and (b) Natural jerk classification; the arsenic occurrence percentage among (c) two classification systems and (d) predicted arsenic vulnerability and arsenic concentration of three districts named Saran, Samastipur, and Vaishali for 2015 and 2019; (e) AUCROC validation curve

for long durations resulting in increased arsenic infiltration into the groundwater (Fig. 1.3c). Arsenic displayed a high correlation with groundwater quality parameters like silicate, bicarbonates, and iron, as per the data obtained from CGWB. The alluvial plain of Bhagalpur, Bhojpur, Buxar, Munger, and Patna have elevated silicate (14–30 ppm), which releases arsenic upon leaching by accelerating the silicate mineral dissolution (Fig. 1.3d). The strong affinity of bicarbonates and pyrites with arsenic also releases it into the groundwater by the formation of complexes on $Fe(OH)^3$ surface sites and by the oxidation of arsenic-bearing sulfide minerals (Fig. 1.3e, f). It is observed that carbon-rich clay beds also facilitate Arsenic dissolution and its release. Elevated arsenic is found in a new alluvial belt along the Ganges (Saha et al. 2009). This contamination is confined to sand, sandy clay, and silty clay layers of 50 m thickness, threatening the drinking water supply of the inhabitants.

1.4.3 Validation

The accuracy of the approach applied was determined by validating the acquired vulnerability with the groundwater arsenic concentrations of three districts named Saran, Samastipur, and Vaishali for 2015 (secondary data) and 2019 (primary data)

(Fig. 1.4d). For these districts, low arsenic was observed in 2015 by CGWB, but the resultant vulnerability map predicted moderate to high arsenic in these districts. The primary data analysis was the same (moderate to high arsenic) for three districts in 2019. Hence, the predicted As vulnerability was validated with the ground situation. Furthermore, the Receiver Operating Characteristics (ROC) curve analysis was computed between sensitivity (True Positive Rate [TPR]) and specificity (1-False Positive Rate [FPR]). This study used the generated arsenic vulnerability dataset and the training dataset of twelve districts to analyze the ROC curve. After creating the ROC curve, the area under the curve (AUC) was calculated to determine the accuracy rate, which ranges between 0.6 and 0.7, indicating the acceptance of the with 60–70% accuracy (Fig. 1.4e). AUC close to 1 is considered perfect accuracy (Hanley and McNeil 1982).

1.5 Conclusion

The current paper endeavored to evaluate groundwater arsenic vulnerability using the Analytical Hierarchy Process (AHP) and Arsenic Vulnerability Index (AVI) in the GIS platform for the districts of Bihar in the vicinity of the Ganga river. AHP was applied for pair-wise comparison considering the difference in the extent of the contribution of each covariate thematic layer: geomorphology, elevation, depth to water level, and the groundwater quality. The study represents the geogenic evidence for groundwater arsenic contamination.

The high rate of silicate dissolution and affinity of pyrites with arsenic releases it into the deep aquifers. Hence, high silicates and iron content in the study area are indicators of groundwater quality, which act as the major factor influencing the arsenic contamination in the groundwater. The high iron dissolution rate from arsenic-bearing minerals of alluvial plains increases the infiltration rate at the low-lying areas accompanying arsenic in the subsurface water. Similarly, the geomorphology and the low elevation of the area also contribute to the elevated arsenic. The consistency of AHP is based on the selected covariates and their assigned weights and rates. In this study, weights were based on literature surveys for each covariate and class of each covariate, which require deep knowledge of the factors regulating the arsenic contamination in the groundwater. The resultant vulnerability map indicated arsenic distribution in a patchy pattern across the state, which can be considered the actual situation on the ground for arsenic contamination. The validation method also confirmed up to 60–70% efficiency. The stakeholders can use this study for planning the mitigation strategy against the elevated arsenic concentrations in the state.

Acknowledgements The authors are grateful to CGWB, Geological Survey of India, United States Geological Sciences, Indian Water Resource Information System, Ministry of Jal Shakti for providing secondary data for smooth research work. Maps throughout the paper created by Software ArcGIS® and ArcMap™ v 10.6.1, intellectual property of Esri, and used under license with Copyright © Esri. The financial support from DST for providing Inspire fellowship (IF180076) to the first author is deeply appreciated.

References

Agarwal R, Garg P (2015) Remote sensing and GIS-based groundwater potential & recharge zones mapping using multi-criteria decision making technique. Water Resour Manage 30:243–260. https://doi.org/10.1007/s11269-015-1159-8

Alam MS, Wu Y, Cheng T (2014) Silicate minerals as a source of arsenic contamination in groundwater. Water Air Soil Pollut 225:2201. https://doi.org/10.1007/s11270-014-2201-9

Bhattacharya P, Chatterjee D, Jacks G (1997) Occurrence of arsenic-contaminated groundwater in alluvial aquifers from Delta Plains, Eastern India: Options for safe drinking water supply. Int J Water Resour Dev 13(1):79–92. https://doi.org/10.1080/07900629749944

BIS (Bureau of Indian Standards) (2012) Specification for drinking water IS 10500: 2012, New Delhi, India. http://cgwb.gov.in/Documents/WQ-standards.pdf

Census of India (2011) Government of India. Provisional population totals, Paper-1 of 2011, Registrar General, India. https://censusindia.gov.in/2011-prov-results/prov_results_paper1_india.html

CGWB (Central Ground Water Board) (2015) Ministry of Water Resources, G. o. I., Groundwater quality scenario. http://cgwb.gov.in/

Chakroborti D, Das B, Murrill TM (2011) Examining India's groundwater quality management. Environ Sci Technol 45(1):27–33. https://doi.org/10.1021/es101695d

Chatterjee S, Moogoui R, Gupta D (2017) Arsenic: source, occurrence, cycle, and detection. In: Gupta D, Chatterjee S (eds) Arsenic contamination in the environment, Springer, Cham. https://doi.org/10.1007/978-3-319-54356-7_2

Das D, Chatterjee A, Mandal BK, Samanta G, Chakraborti D, Chanda B (1995) Arsenic in ground water in six districts of West Bengal, India. Environ Geochem Health 18:5–15. https://doi.org/10.1007/BF01757214

Dhamija S, Joshi H (2022) Prediction of groundwater arsenic hazard employing geostatistical modelling for the Ganga Basin, India. *Water* 14(15):2440. https://doi.org/10.3390/w14152440

GOI (2015) Bihar state profile. http://dcmsme.gov.in/old/dips/state_wise_profile_16-17/Bihar%20%20State%20Profile.pdf

Hanley JA, McNeil BJ (1982) The meaning and use of the area under a receiver operating characteristic (ROC) curve. Radiology 143(1):29–36. https://doi.org/10.1148/radiology.143.1.7063747

Jang Y, Somanna Y, Kim H (2016) Source, distribution, toxicity and remediation of arsenic in the environment – a review. Int J Appl Environ Sci 11(2):559–581. https://www.ripublication.com/ijaes16/ijaesv11n2_19.pdf

Keshavarz-Ghorabaee MA, Maghsoud Zavadskas E, Turskis Z, Antucheviciene J (2018) Simultaneous evaluation of criteria and alternatives (SECA) for multi-criteria decision-making. Informatica, 29. https://doi.org/10.15388/Informatica.2018.167

Kumar CP (2015) Status and mitigation of arsenic contamination in groundwater in India. Int J Earth Environ Sci 1(1):1–10

Lee AHI, Chen WC, Chang CJ (2008) A fuzzy AHP and BSC approach for evaluating performance of IT department in the manufacturing industry in Taiwan. Expert Syst Appl 34(1):96–107. https://doi.org/10.1016/j.eswa.2006.08.022

Levary RR (2008) Using the analytic hierarchy process to rank foreign suppliers based on supply risks. Comput Ind Eng 55(2):535–542. https://doi.org/10.1016/j.cie.2008.01.010

Lynch SD, Reynders AG, Schulze RE (1994) Preparing input data for a national-scale groundwater vulnerability map of southern Africa. Water 20(3):239–246. https://doi.org/10.10520/AJA03784738_1125

Matisoff G, Khourey CJ, Hall JF, Varnes AW, Strain WH (1982) The nature and source of arsenic in Northeastern Ohio ground water. Ground Water 20(4):446–456

Matschullat J (2000) Arsenic in the geosphere – a review. Sci Total Environ 249(1–3):297–312. https://doi.org/10.1016/S0048-9697(99)00524-0

M/o WR, GR RD (2014) Occurrence of high arsenic content in groundwater committee on estimates. http://164.100.60.131/lsscommittee/Estimates/16_Estimates_19.pdf

Neidhardt H, Berner Z, Freikowski D, Biswas A, Majumder S, Winter J, Gallert C, Chatterjee D, Norra S (2014) Organic carbon induced mobilization of iron and manganese in a West Bengal aquifer and the muted response of groundwater arsenic concentrations. Chem Geol 367. https://doi.org/10.1016/j.chemgeo.2013.12.021

Nickson R, McArthur J, Burgess W, Ahmed KM, Ravenscroft P, Rahman M (1998) Arsenic poisoning of Bangladesh groundwater. Nature 395(6700):338. https://doi.org/10.1038/26387

Nordstrom DK (2002) Public health. Worldwide occurrences of arsenic in ground water. Science (New York, N.Y.) 296(5576):2143–2145. https://doi.org/10.1126/science.1072375

Saaty TL (1977) A scaling method for priorities in hierarchical structures. J Math Psychol 15(3):234–281. https://doi.org/10.1016/0022-2496(77)90033-5

Saaty TL (2008) Decision making with the analytic hierarchy process. Int J Serv Sci 1(1):83–98. https://doi.org/10.1504/IJSSCI.2008.017590

Saaty TL, Vargas LG (2006) Decision-making with the analytic network process. Springer Science+ Business Media 251. https://doi.org/10.1007/978-1-4614-7279-7

Saha D, Sarangam SS, Dwivedi SN, Bhartariya KG (2009) Evaluation of hydrogeochemical processes in arsenic contaminated alluvial aquifers in parts of Mid Ganga Basin, Bihar, Eastern India. Environ Earth Sci 1:799–811. https://doi.org/10.1007/s12665-009-0392-y

Smith R, Knight R, Fendorf S (2018) Overpumping leads to California groundwater arsenic threat. Nat Commun 9:2089. https://doi.org/10.1038/s41467-018-04475-3

Triantaphyllou E, Mann SH (1990) An evaluation of the eigenvalue approach for determining the membership values in fuzzy sets. Fuzzy Sets Syst 35:295–301. https://doi.org/10.1016/0165-0114(90)90003-O

Wang S, Mulligan CN (2006) Effect of natural organic matter on arsenic release from soils and sediments into groundwater. Environ Geochem Health 28(3):197–214. https://doi.org/10.1007/s10653-005-9032-y

World Health Organization (WHO) (2022) Guidelines for drinking-water quality: Fourth edition incorporating the first and second addenda. https://www.who.int/publications/i/item/9789240045064

Chapter 2
Livelihood Change and Sustainability Potential in a Sri Lankan Mountain Village

Miyo Matsumoto

Abstract The sustainability of urban areas is the main theme of this volume which is one of the key challenges faced by the global community today. SDG11 (Sustainable Cities and Communities), elobarates the importance of economic, social, and environmental links between urban, peri-urban, and rural areas. However, many rural populations are migrating to urban areas, causing negative effects (e.g., development of slums, poverty, and failure of medical and welfare services) in urban areas and depopulation in rural areas. In mountainous rural areas, where the natural environment is fragile and difficult to live in, people are engaged in considerable efforts to maintain their livelihoods, which could be a factor that pushes people to migrate to cities. By understanding how socioeconomic dynamics affect people's livelihoods in such mountainous rural areas, we can determine how sustainable livelihoods can be developed and finally aim for an appropriate flow of people from rural to urban areas. Thus, this paper aims to clarify how livelihoods have changed in response to socioeconomic changes in a rural mountain village in Sri Lanka. In the studied village, while traditional rice cropping for self-sufficiency remains important, villagers have also sought various sources of cash income, employing multiple livelihood strategies. Although located in mountains, far from lowland urban areas, the studied village had easy access to nearby towns. Therefore, the village can gain income from the non-agricultural sector, which played a major role in their subsistence alongside rice cropping.

Keywords Mountain village · Rice cropping · Non-farm income · Sustainability · Sri Lanka

---

M. Matsumoto (✉)
Kyoto University, Kyoto, Japan

A. Dubey et al. (eds.), *Promoting Sustainability Through Water Management and Climate Change Adaptation*, Advances in Geographical and Environmental Sciences,
https://doi.org/10.1007/978-981-99-5479-7_2

2.1 Introduction

The sustainability of society is arguably the central theme of the international community at present. For example, in 2015, the United Nations adopted the 2030 Agenda for Sustainable Development, which consists of 17 goals. Among them, Goal 11 (Sustainable Cities and Communities) states that the sustainability of urban and rural areas is a problem that cannot be addressed separately, and hence establishing a good connection between urban and rural regions is crucial. This goal aims to control the excessive flow of people from rural to urban areas. Even though rural mountainous areas are mostly regarded as isolated regions with poor access, they have always been affected by global socioeconomic and climate changes. Excess population outflows to urban areas might be one of the reactions to such influences. Mountainous areas have a fragile ecological environment (Watanabe 2004), which is also a harsh environment for humans due to low temperatures and undulating terrain that requires a lot of effort to maintain a living base. Explaining the change in forms of subsistence in mountainous regions in relation to global social changes is an important process to establish sustainable rural livelihoods and to aim for an appropriate relationship between urban and rural areas.

Studies on livelihood strategies in mountainous regions flourished from the 1970s, and the vulnerability of mountainous environments was also noted. To consider the impact of human activities on the environment, comparative studies of mountainous areas were conducted (Guillet 1983). Rhoades and Thompson (1975) proposed that ecologically distinct zones are distributed vertically in mountainous areas, and geological diversity such as slopes and undulations are characteristics of this environment. People have managed to cope with such an environment strategically and have adapted to living in mountainous areas. Kawakita (1977) also stated that mountain land use reflects human adaptation to vertically distributed climatic zones.

Such vertical distribution of land use is the basis for the complex management of multiple livelihood strategies, such as agro-pastoralism which combines livestock raising and farming. Complex livelihoods are considered a system that maximizes productivity while overcoming the harsh natural environment (Mishra et al. 2003). In this paper, the Author considers how land use and livelihood patterns peculiar to mountainous areas are changing due to external socioeconomic dynamics.

2.2 Methods

2.2.1 Studied Site

The studied village P is in the Walapane region, Nuwara Eliya District, Central Province of Sri Lanka (Fig. 2.1). Nuwara Eliya District is located in the central highlands and includes the peak of Sri Lanka's highest mountain, Pidurutalagala

Fig. 2.1 Map of Sri Lanka (*Source* Made by the author based on Vijitha Yapa School Atlas (2017))

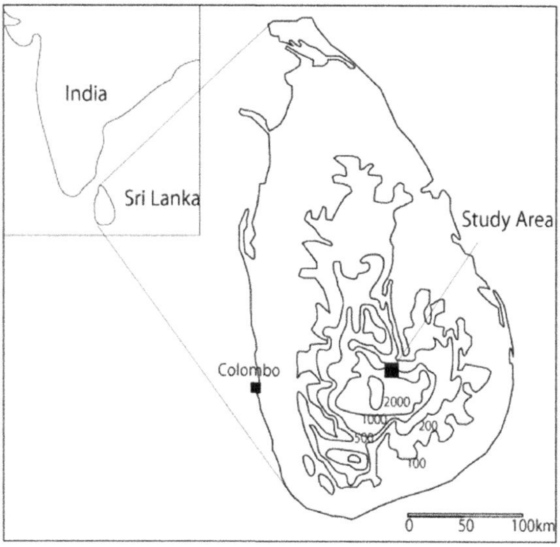

(2,524 m). Sri Lanka is divided into a dry zone with an annual rainfall of 1,000–2,000 mm and a wet zone with 2,000–5,000 mm. The Nuwara Eliya District is located at the boundary of these zones, and its annual rainfall is approximately 1,900 mm.

In the central highlands, tea plantations introduced by the British colonial government during the 1800s coexist with traditional rural villages engaged in rice farming. In the dry zone, rice cultivation has been carried out using a well-developed irrigation system. However, in the central highlands located near the boundary with the wet zone, rainfall is persistent, which is, along with diversion work from rivers, the water source for paddy rice cultivation (Johnson and Scrivenor 1981).

The altitude of village P ranges from 720 m to 1,250 m, and it is located on the east side of the mountain (Fig. 2.2).

At the lowest altitude of 720 m, the river runs northeast, which is connected to the neighboring village on the east side. The boundary on the west side is halfway up the mountain, with an altitude of about 1,000 m, and above this point is a national land. Paved roads run through the village north and south, which are connected to the nearest town N, a 10-min bus ride away.

The scenery of the surveyed village is striking, with rice fields all over the slope. There are two rainy seasons per year, one of which is the harvest season in May (Maha) and the other in November (Yala). This indicates that the villagers conduct double cropping of rice. A waterway was drawn from the mountainside to paddy fields; it was said that the water had never died from before the time of the grandparents (Fig. 2.3). The role of the Agricultural Instructor (AI), who has strong connections with rural Sri Lankans. The AI works with rural villages to achieve stable agricultural income. For example, in the studied village, the AI consulted about the technical aspects of growing cash crops and planned to open a weekly market in town N. He was in charge of an area of about 60 km^2, including 29 Vayama. Vayama is an administrative unit,

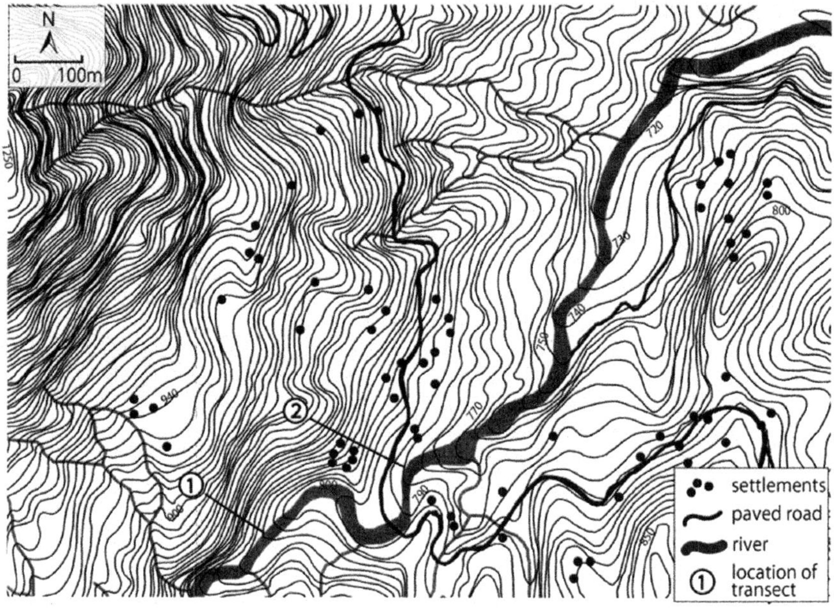

Fig. 2.2 Topography of the Survey Village (*Source* Contour line every 5m. Drawn by the author based on GIS data (62-17), The Survey Department of Sri Lanka)

under the jurisdiction of one administrative officer called the Grama Niladhari (GN), which includes 4–6 natural villages (Shibuya 1985).

Although no one knew the history of village P in detail, there is a place in the region called Radigawela, which is said to have grown the King's rice about 2,500 years ago. Therefore, the place was named after the words Raju ge wela (King's paddy). In addition, in the war with the British colonial government, the local men were forced to become soldiers because they were said to be healthy due to the rice they were eating as a staple. Therefore, it seems that the survey area was established as a rice-cropping village at least before colonization (Interviews with AI, August 6, 2019).

2.3 Survey Method

A field survey was carried out in Sri Lanka from the 1st to the 21st of August 2019. During my stay in Colombo, the Author obtained a topographic map and ArcGIS data at the Survey Department of Sri Lanka. During my stay in the rural village, I conducted the following two surveys.

Fig. 2.3 Paddy fields of village P (*Source* Author (August 7, 2019))

2.3.1 Survey of the Village's Topography

The entire village is located on the mountain slopes. To clarify the relationship between the slopes and residents' land use, Author has set two transect lines (refer to ① and ② on Fig. 2.2) and drew a cross-sectional map of the terrain using a hand level and a distance-measuring instrument. Current land use was plotted in the drawn map to determine the relationship between slope and land use.

2.3.2 Interview Survey

Interviews that were focused on subsistence activities were conducted with 19 householders, among the 84 households in village P. Topics included family structures, number of migrant workers in the family, number of family members employed full-time, area of paddy and other fields, crop planted in the reference year, amount of income obtained from the crop in 1 year until the harvest of May 2019, and other means of cash income.

Interviews with officials were also carried out. Author asked Grama Niladhari (G.N.) who has jurisdiction over village P about the population of the village and the status of subsidy payments. Interviews with Agricultural Instructor (A.I.) concerned macro policies related to the region.

2.4 Results

2.4.1 Topography of the Village

There were various angles of slopes in the village; land use was investigated in relation to slope surfaces.

Figure 2.4 shows a cross-sectional view of Transects 1 and 2. Transect 1 is located on the western slope of the village and covers the range from 810 m above sea level at the bottom of the valley where the river flows, to 880 m above sea level on the mountainside. Its horizontal distance is approximately 100 m. Starting from the valley bottom, there is a gentle slope going up a horizontal distance of approximately 20 m (hereinafter, the horizontal distance indicates the distance from the starting point). The average angle of this sloping surface is 19°, and it is used as a paddy field. Thereafter, it becomes a steep cliff up to a horizontal distance of 90 m and rises about 60 m in relative height. The average angle of this slope is 42°. This slope was once used for tobacco growing, but it has been abandoned for a long time. For the rest of the distance up to a horizontal distance of 100 m, the slope becomes gentle again, with an average angle of 21°. This sloped surface is used as a paddy field.

A cross-sectional view of Transect 2 is similar to Transect 1 It is located on the western slope of the village and covers the range from 785 m above sea level at the bottom of the valley where the river flows, to 845 m above sea level on the mountainside. The horizontal distance of the line is approximately 160 m. Starting from the bottom of the valley, there is a gentle slope up to a horizontal distance of 30 m. The average angle is 20°, and land use is primarily for paddy fields. Up to a

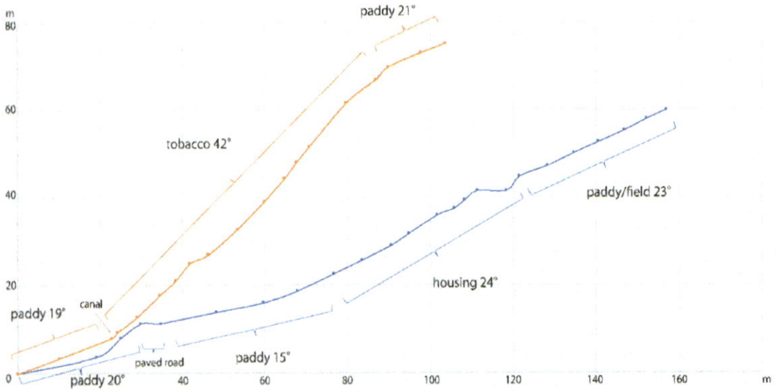

Fig. 2.4 Cross section of Transect 2 (*Source* Author's field survey (August 2019))

Table 2.1 Inclination of slope and land use

Inclination of slope surface	Land use
42°	Fallow
24°	House, home garden
~23°	Paddy field, vegetable field

Source Author's field survey (August 2019)

horizontal distance of 80 m, the slope is even gentler, with an average angle of 15°. This sloped surface is also used for paddy fields. Subsequently, the slope becomes slightly steeper up to a horizontal distance of I 120 m, with an average angle of 24°. This sloped surface is used for residences and home gardens. The end of transect has an average angle of 23°; hence, it may be regarded as an extension of the residential slope surface in terms of its angle. This surface is used as paddy fields or vegetable fields. As the slope continues to the west, the altitude rises and continues to the top of the mountain. However, the area above this is bushy where many wild monkeys live, so further surveys were abandoned.

Summarizing the relationship between topography and land use (Table 2.1), a steep slope surface of 42° is currently abandoned, and a sloped surface of 24° is used for residences. In addition, some households create gardens around their residences and cultivate pepper and capsicum. The terrain surface with the gentlest slope up to 23° is used as paddy or vegetable fields.

The slope of the land alone never determines the land use, and other factors, such as access to water, soil, and relationships within the village, may affect the land use. However, this study does not focus on the process of land use formation, but functions as an index of how land use has changed. Therefore, it is useful to clarify the relationship between land use and slopes, which are the natural conditions peculiar to mountainous areas.

2.4.2 *Livelihood Strategies*

From the interviews, it was observed that the different livelihood strategies and how they relate to vertical land use. This includes rice paddy and vegetable cultivation, which are the main agricultural activities of village P additionally; sources of cash income are explained.

2.4.2.1 Paddy Rice Cultivation and Cash Income

Apart from one (household no. 14) of the 19 households interviewed, all households were involved in rice cultivation.

Figure 2.5 shows a graph of the yield of rice in each household and the amount of rice and vegetables sold during the year up to May 2019. As the surveyed village

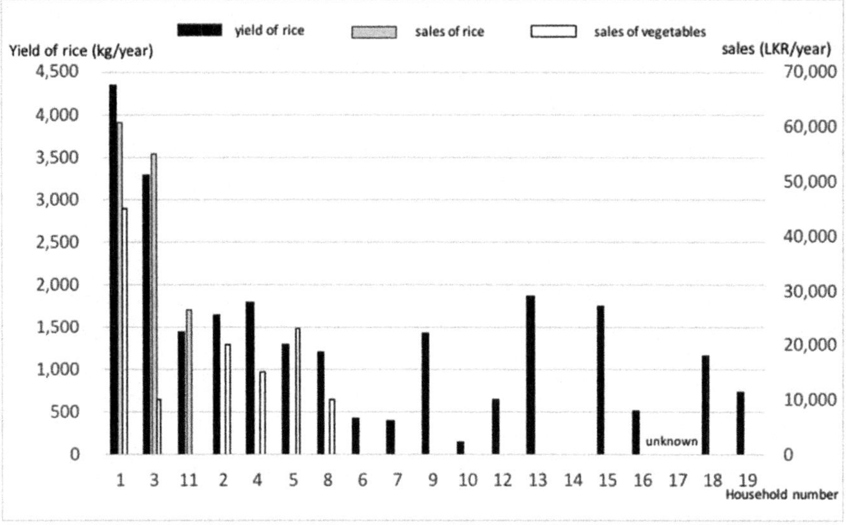

Fig. 2.5 Yield of rice and sales of rice and vegetables (*Source* Author's field survey (August 2019))

carries out double cropping of rice, both annual harvests (May and November) were included. When members of all households were asked about the use of harvested rice, they answered that it was for personal consumption. Among these, 14 households answered that their rice was primarily for personal consumption, and 3 answered that they sold a large amount of rice in addition to their personal consumption.

Household no. 1 has six family members: grandparents, head of the household, wife, and son and daughter. The head of the household owns the largest paddy field in the village (5 acres) and a 3.5-acre field. He sold 1,744 kg of rice out of a 4,360 kg yield and gained 60,800 Sri Lankan Rupees (LKR) which was the equivalent of 342 US$ on 2019 August 1st. The remaining rice is for the family's consumption for the entire year. In addition to selling rice, his wife works at a bank, and his son is a police officer. In addition, his father receives a monthly 1,900 LKR old-age pension (for those aged over 70). The income earned from the sale of rice is mainly used for the purchase of farm equipment and his daughter's education. He also said that he was saving cash in case of a poor harvest or in case he could not work. He also grows tomatoes, cabbage, and eggplant; about 80% of the harvest is sold, which makes 35,000 LKR. Food expenses are mainly covered by his wife's income.

Household no. 3 has five family members: the head of the household, his wife, son, and two daughters. The head of the household owns a 2-acre paddy field and cultivates another 2.5 acres through sharecropping. In some years, he changed half an acre of paddy fields into vegetable fields. Out of his 3,300 kg yield of rice, 990 kg was sold and gained 55,000 LKR. The sale price achieved by household no. 1 was 35 LKR/kg, whereas it was 56 LKR/kg for household no. 3.This is because the rice cultivar Samba, which household no. 3 grows, is more expensive. Income is mainly consumed for labor costs for farm work and farm equipment purchases, as well as education costs and living expenses. Moreover, cabbage, green beans, and tomatoes were sold to make 4,000 LKR, but there was no profit. This family does not have a non-agricultural income.

Household no. 1 is investing in large paddy fields in anticipation of harvesting a surplus in addition to that required for personal consumption, and it has two family members who are regularly employed in the non-agricultural sector. Therefore, there is a margin for labor costs and funds for agricultural investment, such as pesticides. Household no. 3 does not have any non-agricultural income, but it seems that they are strategically cultivating rice to increase their income, such as by cultivating more highly priced varieties. Both households position rice cultivation as a source of cash income and invest money in it.

Although Fig. 2.5 does not reflect this in the rice sales, four households answered that they occasionally sell rice when a large amount of money is needed, such as for farm medicines, schools, and weddings. They only sell the amount required to raise the money they need. As rice is bought in town N all year round, rice plays a role similar to insurance that can be converted into cash if sudden expenses emerge, even for households that do not generally sell rice.

In addition, households that do not own paddy fields secure a rice harvest by sharecropping, which accounts for 12 out of 19 households, indicating a shortage of paddy fields. In rice cultivation, the mutual aid labor exchange called attam is practiced during tilling of paddy fields, the building of ridges, transplanting, weeding, and harvesting. If someone has asked for help to cultivate rice by attam, he is obliged to return the favor. In the studied village, rice cultivation is still carried out as an important livelihood strategy, both for personal consumption and as an emergency cash source, and this strategy is supported by traditional activities, such as sharecropping and labor exchange.

2.4.2.2 Vegetable Cultivation and Cash Income

In response to the question about what they do with their harvested vegetables, all six households that cultivated vegetables said they sell them. Among them, two were the households mentioned above (1 and 3), who also sell a large amount of rice. Here, three other households are listed as examples.

Household no. 4 has five family members: the head of the household, his wife, son, and two daughters. The father owns a 2.5-acre paddy field together with his other three brothers. In 2019, their annual yield was 1,800 kg, which they divided equally among themselves. The head of this household does not sell his quota. He also cultivated 0.5 acres of land owned by the temple, where he grew green beans, cabbage, and tomatoes, which sold for a total of 14,000 LKR. Half of this amount will be donated to the temple, so as to do well. Most of the farm work is done by the wife, with help from the attam, because the head of the household works at a post office in town N, which gains a monthly income of 35,000 LKR. His children are sent to elementary schools in town N by bus, instead of the neighborhood elementary schools that the children in the village typically attend. Both schools are public, which implies that tuition is free. However, both parents in household no. 4 think that the school in the town has a higher level of education.

Household no. 5 has four family members: the head of the household, his wife, and two sons. The only land the head of household owns is a home garden; he does not own any paddy or other fields. He cultivates a 1-acre paddy field through a sharecropping arrangement. In 2019, his quota was 1,300 kg, which was all for personal consumption. In addition to the paddy fields, he rented a 0.5-acre field for 10,000 LKR/year and cultivated tomatoes, cabbage, and green beans, which gained a total of 23,000 LKR. He owns nine buffaloes, which help him perform farm works such as tilling of paddy and threshing. He also earns income by doing work for other villagers with his buffaloes. By working on up to about 0.25 acres of paddy, he earns about 1,500 LKR/day, and in the peak season, he will be hired by the villagers about 15 days a month. In fact, during the interviews, at least seven households asked household no. 5 to do farm work as wage labor. The family does not have a non-agricultural income.

Household no. 8 has four family members: the head of household, his wife, son, and daughter. The head of household owns 1.5 acres of paddy fields, of which 0.5 acres may be used as a field in some instances. In 2019, 1,210 kg of rice were harvested. He also harvested tomatoes, cabbage, green beans, and capsicum, which made 10,000 LKR. He and his wife sell lunch boxes in the office of the Ministry of Agriculture in town N, using the rice and vegetables they have grown. Their monthly income, after subtracting the rent, is about 50,000 LKR, from which their costs of living and agriculture are paid. In the peak farming season, the husband will do the farm work or, in some cases, ask the villagers to help as a part-time job (including household no. 5).

Selling vegetables generally requires the involvement of middlemen. For example, for the past 20 years, vegetables have been bought by the broker Mr. S. He comes to the village by truck and stops by houses where there are bags full of vegetables on the side of the road. Vegetables are bought and sold by at least two more middlemen and are finally sold at the market of Dam bulla, 90 km north of the village. The price of tomatoes is 20 LKR/kg in the village, but they are sold at 100 LKR/kg in the city market.

In the households mentioned above, 80–100% of the harvested vegetables were sold, making from 10,000 LKR to 23,000 LKR. The monthly salary of the post office staff is 35,000 LKR, so vegetable sales cannot be the main income of the household, but may contribute to toiling needs on a temporary basis. The head of household no. 4 was a post office employee, the head of household no. 5 was an agricultural wage laborer, and the parents in household no. 8 ran a lunch box shop. Each household had other major sources of cash income.

2.4.2.3 Non-agricultural Cash Income

There are 13 households that do not grow vegetables. All of them cited as their reasons for not doing so, "the monkeys eat them" and "the purchase price is too low." Although not among the 19 households surveyed, some villagers lamented that they invested 300,000 LKR to grow pumpkins but their sales value did not even reach 50,000 LKR. In addition, the head of household no. 3 said that his vegetable cultivation was in the red in 2019. In addition, the head of household no. 17 said that the purchase price of cabbage was as low as 2 LKR/kg, so he consumed what he grew himself or gave it to other people in the neighborhood. Even for those who cultivate vegetables in the expectation of income, vegetable prices fluctuate significantly.

Here, Author will look at three of the households that do not grow vegetables and examine their livelihoods.

Household no. 9 has four family members: the head of household, his wife, son, and daughter. The head of the household does not own any paddy or other fields, but he cultivates 1.5 acres of paddy through sharecropping. In 2019, his annual quota was 1,435 kg, which was all for personal consumption. Similar to household no. 4, he works at the post office and earns a monthly salary of 35,000 LKR which was equivalent to 197 US$ on 2019 August 1st. The wife also has a part-time farming job. However, during the peak season of farm work, the head of the household also works in the paddy part-time, usually on his days off from the post office. He owns a cultivator, which allows the household to independently perform several agricultural activities, such as tilling and harvesting. This agricultural work is usually done in the form of attam, the abovementioned mutual aid among villagers. However, participating in attam is difficult for the heads of households who have time constraints due to working at the post office, because they do not have any time to return the labor that is offered to them. Villagers who have a cultivator are able to manage all the processes within their household, and as they do not participate in mutual aid, they are able to obtain part-time work when helping with other people's farm work.

Household no. 13 has five family members: grandmother, head of household, wife, daughter, and son. The head of household owns 1 acre of paddy and consumes 1,870 kg of rice harvest within the household. He runs a clothing store in town N and earns 20,000 LKR monthly. He said that he did not grow vegetables for cash income because of the damage caused by animals. He said that investment in buying fences and nets to protect vegetables from animals is costly and those who have such funds do not need to grow vegetables in the first place. Animal damages to the paddy are caused by wild boars digging their shores, causing water to flow out. Therefore, he goes to his paddy every morning and evening to make sure there is no damage. However, damage caused by animals is less serious in rice than in vegetable cultivation.

Household no. 15 has four family members: the head of household, his wife, son, and daughter. The head of household owns 1 acre of paddy, and generally, all of the rice harvests of 1,760 kg are consumed inside the household, but they sometimes sell about 100 kg to buy fertilizers and pesticides. In the summer of 2019, 100 kg of rice was sold for 4,250 LKR. The head of the household is a carpenter and earns about 30,000 LKR monthly. He has also opened a general store on the first floor of his house, and its profit is about 24,000 per month. He says he does not grow vegetables because he is busy with his carpentry job and his store.

As mentioned above, the interviews revealed that some households do not grow vegetables as their source of cash income because of the many damages caused by animals. Even if they put all their efforts into growing vegetables, such as by investing in animal defenses and hiring a labor force for farm work, the price of vegetables fluctuates and therefore the income is not stable. In addition, there is non-agricultural income in many households, which creates time constraints if one was engaged in an independent employment and sales business, so they do not have time for farm work. Furthermore, as they earn a salary from their other jobs, they do not feel the need to cultivate vegetables.

2.4.2.4 Households That Grow Neither Rice nor Vegetables

Household no. 14 is an old couple who owns no paddy or other fields. They run a stall where they sell some snacks. The stall is located beside their house and makes about 10,000 LKR a month. Once a month, the husband goes to Kurunegala, a city 60 km north of the village, by bus and buys tobacco leaves to sell in town N. He said that the previous month's profit was 4,000 LKR. Moreover, they receive old-age pensions and subsidies for poor a household, which gains a total of 3,400 LKR. They also receive allowances from their two married daughters. During the interview, their stall and house were for sale: they were planning to move to their daughter's house as soon as the properties were sold.

2.4.2.5 Migrant Workers

Thus far, the author has mentioned about cash incomes within the village or the nearby town. However, households with family members who stay and work outside the village were few. In addition to the households who are grown up children have left their home and live in a large city such as Colombo, there were two households who answered that the husband goes to large towns to work. In household no. 18, the husband occasionally goes to Colombo for carpentry works, and in household no. 19, the husband goes to Colombo once in 2 weeks also for carpentry works. Migrant works often refer to a work wherein the head of the household moves away from his family and makes money to either bring back home once in a while or send it through remittances. In this context, migrant work of village P seems not a main subsistence.

As mentioned above, interviews were conducted mainly on the cultivation of rice and vegetables and about the sources of cash income. Table 2.2 describes the sources of cash income, including non-agricultural sources. There are roughly three groups of households. Group 1 sells a large amount of rice, invests well, and grows rice to gain cash income. Group 2 grows vegetables despite the risk of deficits, but has an additional source of non-vegetable-based cash income. Group 3 does not grow vegetables at all and focuses their time and effort on other sources of cash income. This group, who separate agriculture from cash income sources, is the largest group. This indicates that the non-agricultural sector occupies a large part of

Table 2.2 Sources of cash income

Group	Household number	Agriculture			Non-agriculture			Migrant work	Allowance (poor/aged)
		Rice	Vegetable	Wage labor	Salary	Self-owned business	Wage labor		
1	1	○	○		○				○
	3	○	○						○
	11	○		○					
2	2		○						○
	4		○		○				
	5		○	○					○
	8		○			○			
3	6						○		○
	7								○
	9			○	○				
	10								○
	12						○		○
	13					○			○
	14					○			○
	15					○	○		○
	16				○				○
	17						○		○
	18			○	○		○	△	○
	19					○	○	○	○

Source Author's field survey (August 2019)

the village's livelihood. However, regarding migrant work, only one household was involved regularly.

2.4.3 Changes in Livelihood

Noticeably, the primary livelihood strategy in the surveyed village is currently rice cultivation for personal consumption, and there is a large amount of non-agricultural cash income. However, it was said that tobacco cultivation was popular in the 1980s, and some households grew tobacco even in paddy fields on gentle slopes. Here, the outline of land use and tobacco cultivation and the process leading to the ban on tobacco cultivation will be explained.

The tobacco market was virtually monopolized by the Ceylon Tobacco Company, and they purchased tobacco leaves from contract farmers (Gunati lake 1995). However, soil erosion due to tobacco cultivation is particularly problematic in parts of the Central Highlands, including the study village area (Ministry of Environment and Renewable Energy, Sri Lanka 2014). Tobacco takes up more nitrogen, phosphorus, and potassium than other major crops, so it can deplete soil nutrients (Novotny et al. 2015), causing extreme soil degradation. It is said that 70 tons/ha/year of soil erosion occurs from tobacco fields in mountain areas where no mitigation measures are taken (Ministry of Environment and Renewable Energy, Sri Lanka 2014). Hanguranketha and Walapane, where soil erosion is particularly severe, are in the upper reaches of the Mahaweli River, and the eroded soil was deposited in the Victoria Reservoir that connects to the Mahaweli River. This was an obstacle to the highly anticipated Mahaweli Development Programme, which was formulated by the Sri Lankan government in the 1960s with the support of UNDP and FAO. It was a multipurpose regional development program focused on the development of water sources in the Mahaweli River system and the irrigation development of the region. Given this situation, tobacco cultivation has diminished since the late 1980s.

The following quote is from an interview with Mr. H, who was involved in the area as the GN during the peak of tobacco cultivation. Tobacco cultivation began in 1965, and during the 1970s and 1980s, almost all farmers in the village grew tobacco, and households that did not grow it were involved in it through wage labor. Rice was cultivated for self-consumption and tobacco was a source of cash income, which made a lot of money. The tobacco company had instructed them to create shores to prevent soil erosion. However, there was a limit to the prevention of soil erosion on slopes, and tobacco cultivation was restricted in 1988. After that, vegetables were grown as cash crops instead of tobacco, but it was not profitable. Thirty years ago, the only option was to cultivate the land, but now you can make money by going to the town (Interview; August 11, 2019).

Table 2.3 shows changes in land use. In the 1980s, when tobacco cultivation was at its peak, rice cultivation for personal consumption was carried out on gentle slopes, and tobacco was mostly cultivated on steep slopes. However, some households stopped cultivating paddy altogether, and started cultivating tobacco on the gentle slopes. Since 2019, tobacco fields have disappeared from both gentle and steep slopes, with gentle slopes being used for paddy and other fields, and steep slopes being left fallow.

Aso (2009) conducted a survey in mountain villages in Japan and found that land use in rural areas has generally changed to residential land or parking lots or has been abandoned, as livelihood strategies shift from agricultural to non-agricultural.

Table 2.3 Changes of land use in the surveyed village

Percentage	1980s	2019
42°	Tobacco field	Fallow
24°	House	House, home garden
~23°	Tobacco field, paddy field	Paddy field, vegetable field

Source Author's field survey (August 2019)

As observed in the survey village, the extent of abandoned land increases as non-agricultural livelihoods increase.

2.5 Discussion

In the mountainous regions of Sri Lanka, which have been engaged in agriculture focused on rice cultivation since before colonization, there was a booming economy of tobacco cultivation (including village P) that started in 1965. The 1980s were the heyday of tobacco cultivation, with steep slopes used as tobacco fields and gentle slopes as paddy fields. By using a combination of terrain surfaces with different slopes, farmers could have both self-sufficient crops (paddy fields) and cash income sources (tobacco fields). As mentioned initially, land use in mountainous areas has been adapted to each distinct vertically distributed ecological environment. This vertical division is mainly characterized by differences in climate, but since the difference in elevation within the survey village was small, it was not considered. However, the undulating terrain, which is a combination of sloped surfaces of various inclinations, is also a natural environment peculiar to mountainous areas. Therefore, the vertical land use pattern of the survey village can be considered representative of mountainous areas characterized by multiple livelihood strategies, which makes maximum use of the mountain's restrictive natural environment.

Eight of the 19 households interviewed lived in the village and went to the town to earn a fixed income, and the other two conducted their business beside the road in the village. Permanent migrant worker was only one, indicating that villagers' earnings mostly derive from either within the village or nearby town. In contrast to vegetable cultivation, animal damage does not cause as many problems for rice cultivation and rice can be sold for cash. Households that can afford to invest cultivate rice with the expectation of greater profits, and even households that cannot invest very much can still regard rice cultivation as insurance in the event of an emergency. Therefore, in village P, while all households maintain rice cultivation for personal consumption, there are three distinct groups depending on the source of their cash income:

1. A group that grows rice on the premise of cash income.
2. A group that attempts to grow cash crops while securing cash income sources, including from a non-agricultural sector.
3. A group that focuses mainly on the non-agricultural sector as a cash income source.

The fact that there are no longer organic links among types of land use is a result of residents expecting more productive livelihoods from sources other than the natural environment. In the rural areas of third world countries, people become less dependent on natural resources and more dependent on wage labor (Hajdu 2005). The village for survey is no exception, and the people that live there take advantage of its accessibility to the town to change their livelihood strategies. However, rice cultivation remains an important part of their livelihood, as a source of personal consumption crops and as cash income for emergency. The residents have changed their source of cash income in response to socioeconomic changes while retaining traditional activities that utilize the natural environment. Such type of change is possible due to the advantage village P has, which is the good access to the neighboring town.

2.6 Conclusion

Although mountainous areas seem to be isolated compared to urban areas, they are closely linked to macro socioeconomic activities, and livelihoods there change in response to these changes. In this process, organic links between land surfaces, characteristic of mountainous rural areas, are lost, and multiple livelihood strategies combining both agricultural and non-agricultural work are practiced. In this paper, it seemed that village P has gone along the transition with a good condition because it has good access to the neighboring town. However, to explain urban–rural relationship in detail, the livelihood changes in villages with poor access or those areas that are too high to cultivate rice should also be investigated, and this remains an issue for further studies.

Acknowledgements This work was supported by JSPS KAKENHI (Grant Number 17J40062). I thank JSPS for their support.

References

Aso M (2009) Transformation of spatial composition in rural village and problems of how to conserve the traditional landscape. J Architect Plan Environ Eng 74(646):2637–2645 (in Japanese)

Guillet D (1983) Toward a cultural ecology of mountains: the Central Andes and the Himalaya compared. Curr Anthropol 24(5):561–574

Gunatilake HM (1995) An institutional analysis of soil erosion problem of hill country tobacco cultivations in Sri Lanka. In: Hundekar SG (ed) Management of the rural sector. Mittal Publications, pp 78–90

Hajdu F (2005) Relying on jobs instead of the environment? Patterns of local securities in rural Eastern Cape, South Africa. Social Dyn 31(1):235–260

Johnson BLC, Scrivenor MM (1981) Sri Lanka: land, people and economy. Heinemann Educational Books

Kawakita J (1977) Vertical structure of various cultures in central Nepal and the Himalayas. Quart Anthropol 8(1):3–80 (in Japanese)

Ministry of Environment and Renewable Energy, Sri Lanka (2014) National Action Program (NAP) for combating land degradation in Sri Lanka 2015–2024. Karunaratne & Sons

Mishra C, Prins HHT, van Wieren SE (2003) Diversity, risk mediation, and change in a Trans-Himalayan agropastoral system. Human Ecology 31(4):595–609

Novotny TE, Bialous SA, Burt L, Curtis C, Costa VL, Iqtidar SU, Liu Y, Pujari S, Tursand'Espaignet E (2015) The environmental and health impacts of tobacco agriculture, cigarette manufacture, and consumption. Bull World Health Organ 93:877–880

Rhoades RE, Thompson SI (1975) Adaptive strategies in alpine environments: beyond ecological particularism. Am Ethnol 2(3):535–551

Shibuya T (1985) "Sokari", a ritual drama of Sri Lanka: an interpretation of the symbolism. Shien 44(1):54–16 (in Japanese)

Vijitha Yapa Publications (2017) School atlas: latest maps on Sri Lanka and the world. Vijitha Yapa Publications

Watanabe T (2004) Fragility of mountain geo-ecosystem and present situation and agendas of geo-ecological studies. J Geogr 113(2):180–190 (in Japanese)

Chapter 3
Ensuring Potable Water Supply to Rural Areas: A Case of Rural Water Supply in Tumkuru District, Karnataka, India

Jisa Shaji, Shivam Sakshi, and Gopal Naik

Abstract Access to clean water is a major challenge across many Indian rural regions. Although the local governments in rural Karnataka are striving to provide water to the households through strategically located borewells, the quality of this supplied water is a major concern. In many regions, water is consumed with little to no treatment of the supplied bore well water. This study focuses on developing a system for accessing information about the portability of underground water by various stakeholders. For this study, we have selected two Gram Panchayats (GPs) of Madhugiri taluk in the Tumkuru district of Karnataka. The GPs are situated in eastern Karnataka, which is a drought-prone, rocky region. We examine the key concerns of how water is consumed in the region and people's reluctance toward the use of public RO filters. Based on our findings, we propose a framework to ensure regular water monitoring by the concerned authorities and dissemination of information to the relevant stakeholders, especially consumers, through a common digitized report providing information about the quality of drinking water. This will help consumers make well-informed decisions regarding their drinking water source(s), as well as improve the functioning of the water regulatory system.

Keywords Water quality · Rural population · Groundwater · Quality assessment · Right to information

---

J. Shaji (✉)
Economics and Social Science Area, Indian Institute of Management, Bangalore, India
e-mail: jisa.shaji@iimb.ac.in

S. Sakshi · G. Naik
Indian Institute of Management, Bangalore, India

A. Dubey et al. (eds.), *Promoting Sustainability Through Water Management and Climate Change Adaptation*, Advances in Geographical and Environmental Sciences,
https://doi.org/10.1007/978-981-99-5479-7_3

3.1 Introduction

Water is an integrated natural resource that sustains life on earth. Access to clean and safe drinking water is a necessity of human life and essential for survival and development. Nearly 76 million people in India do not have access to safe and clean drinking water (Economic Times 2016). Increasing population and climate variation, such as severe droughts, are leading to rapid changes in the demand and supply of water in the country (Misra 2014). Nearly 80% of India's 1.38 billion people depend on groundwater for their domestic, agricultural, and industrial requirements (Central Ground Water Board 2021). In India, as of April 2020, more than 28 million people are affected by groundwater contamination with iron, arsenic, and fluoride as some of the chief contaminants (Jaganmohan 2021).

Karnataka is the second most drought-prone state in India, and groundwater is a primary source of drinking water. The state is struggling to meet the drinking water requirements, particularly in rural areas as several taluks of the state are perennially drought-affected and experiencing depleting groundwater levels. The Karnataka government is constantly developing various initiatives and schemes to make "Karnataka—a water-secure state" by providing regular and sustainable clean drinking water to all households. The Rural Drinking Water and Sanitation Department (RDWSD) aims to provide proper sanitation facilities and clean drinking water in the rural areas of Karnataka (*RDWSD Karnataka—Official Website* 2021). In this study, we focus on groundwater quality assessment of two selected gram panchayats (GPs) in Madhugiri taluk of Karnataka. The study examines the key concerns in the region and suggests a possible framework to improve the situation.

3.2 Background of the Study

Surface and groundwater are the primary water resources of Karnataka. The availability of Karnataka's net groundwater resources is estimated as 15.3 billion cubic meters (BCM) and the surface water is approximately 7663 thousand million cubic feet per annum (TMC/a), of which only 45% can be economically utilized(Water Resources 2021). Figure 3.1. shows the surface and groundwater availability in Karnataka in the seven river basins and Fig. 3.2 shows the water resources of Karnataka at a glance.

The increase in the demand for water has tremendously increased the dependency on groundwater resources in several districts of the state. This is evident from the significant increase in the number of borewells and tube wells in the state (Raghavan et al. 2017). Owing to overutilization and exploitation of water for construction, irrigation, and other industrial activities, the water quality has deteriorated and is not potable in many places in India. Groundwater quality assessment reports in Tirupur (Tamil Nadu), Nalgonda (Andhra Pradesh), and Kakinada (Andhra Pradesh) have shown that the concentration of contaminants such as iron, fluoride, nitrate,

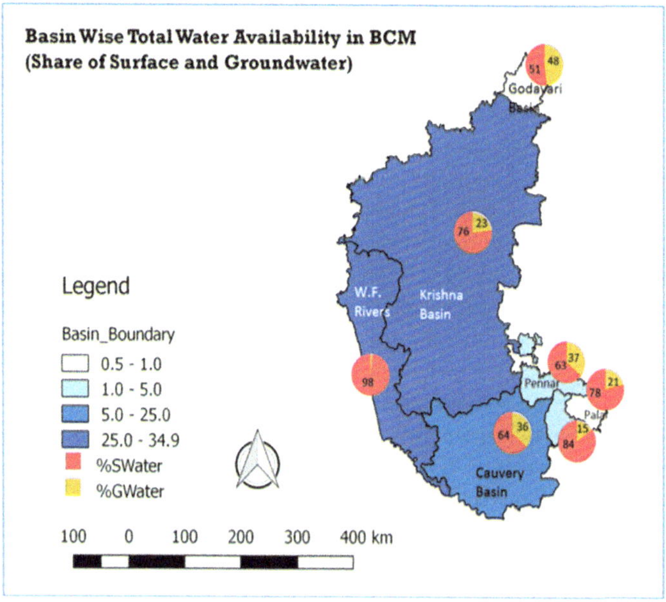

Fig. 3.1 Basin wise total water availability in BCM (*Source* [Raghavan et al. 2017])

Water balance		Sectoral demand (2000)	
Rain and other gains	236 billion m³/a	Household	1.65 billion m³/a
Evapo-transpiration	110 billion m³/a	Agriculture (irrigation)	31.43 billion m³/a
Releases to sea and other states	109 billion m³/a	Industry	1.35 billion m³/a
Input over output	18 billion m³/a	Power sector	1.50 billion m³/a
Surface water		Others	1.50 billion m³/a
Economically available	98 billion m³/a	**Irrigation**	
Present utilisation	48 billion m³/a	Water consumed	31.43 billion m³/a
Groundwater		Groundwater thereof (absolute)	9.7 billion m³/a
Replenishable	15.9 billion m³/a	Groundwater use by quantity	31%
Available	15.3 billion m³/a	Groundwater use by net area	37%
Present utilisation (absolute)	10.7 billion m³/a	Gross irrigated area	41.0 *lakh* ha
Present utilisation (relative)	70%	Net irrigated area	33.9 *lakh* ha
Irrigation thereof	9.7 billion m³/a	Gross irrigated area as share of cultivable area	32%
Domestic, industry thereof	0.97 billion m³/a		

Fig. 3.2 Water Resources of Karnataka (*Source* Water Resources [2021])

chloride, and heavy metals in the samples was beyond the permissible limits. Thus, proper actions are needed to reduce pollution, monitor water quality at the source, and pretreatment of the water for domestic purposes (Ramaswami and Rajaguru 1991; Ravichandran and Pundarikanthan 1991; Govardhann 1990; Davithuraj and Manjunatha 2014).

Karnataka is among the top three states, along with Rajasthan and Tamil Nadu, with high groundwater pollution. Large concentrations of fluoride, nitrate, and lead have been found in groundwater in three districts of Tamil Nadu (Kulkarni 2019).

Reports from the Department of Rural Drinking Water and Sanitation, Karnataka, highlight that groundwater samples drawn from over 25 districts of Karnataka showed high concentrations of fluoride (above 1.5 parts per million (ppm)), nitrates (above 45 ppm), and high total dissolved solids (electrical conductivity (EC) above 3000 μS/cm), the crisis is accentuated by the over-exploitation and improper management of industrial effluents. According to the 2020 assessment, the groundwater of 97 taluks of Karnataka has been categorized under the over-exploited, Critical, and Semi-critical (OCS) categories owing to the presence of several contaminants and pesticides beyond the permissible limit. Figure 3.3 shows the permissible limits of the contaminants in drinking water quality as per the Bureau of Indian Standards (BIS: 10500 2012) and Fig. 3.4 shows the international permissible limits of the constituents in drinking water. The sources and potential health hazards of the contaminants in the groundwater are shown in Tables 3.2 to 3.5 in the Appendix.

Figure 3.3 shows that the permissible limits for fluorides, iron, and arsenic are low, particularly for arsenic. High amounts of arsenic can cause acute and chronic toxicity, liver and kidney damage, decreased blood hemoglobin, and cancer. Excess fluoride content in water can cause crippling bone disorders. The World Health Organizations' (WHO) permissible limits are lower than ICMR-prescribed limits on many parameters (Fig. 3.4.

RURAL DRINKING WATER & SANITATION DEPARTMENT

STANDARD OPERATING PROCEDURE

Issue No.& Date: 01 & 09.09.2019	Revision No. & Date: Nil
SOP No.: RDWSD/AL-PL/01	Title: Acceptable limit/Permissible limit for Drinking water

BIS Specifications 10500:2012

PARAMETERS	ACCEPTABLE LIMITS	PERMISSIBLE LIMITS
pH	6.5-8.5	No relaxation
Turbidity, NTU	1	5
Total Dissolved Solids, mg/L	500	2000
Calcium(as Ca), mg/L	75	200
Chloride(as Cl), mg/L	250	1000
Fluoride(as F), mg/L	1	1.5
Iron(as Fe), mg/L	0.3	1.0
Arsenic (as As), mg/L	0.01	0.05
Magnesium(as Mg), mg/L	30	100
Nitrate(as NO3), mg/L	45	No relaxation
Sulphate(as SO4), mg/L	200	400
Total Alkalinity, mg/L	200	600
Total Hardness, mg/L	200	600
Total Coliform Bacteria, MPN	Absent	-
E.coli, MPN	Absent	-

Fig. 3.3 Water quality parameters and BIS sandards for various constituents (*Source* BIS: 10500 [2012])

Parameters	USEPA	WHO	ISI	ICMR	CPCB
pH (mg/l)	6.5-8.5	6.5-8.5	6.5-8.5	6.5-9.2	6.5-8.5
Turbidity NTU	-	-	10	25	10
Conductivity (mg/l)	-	-	-	-	2000
Alkalinity (mg/l)	-	-	-	-	600
Total hardness (mg/l)	-	500	300	600	600
Iron *mg/l)	-	0.1	0.3	1.0	1.0
Chlorides (mg/l)	250	200	250	1000	1000
Nitrate (mg/l)	-	-	45	100	100
Sulfate (mg/l)	-	-	150	400	400
Residual (mg/l) free Chlorine	-	-	0.2	-	-
Calcium (mg/l)	-	75	75	200	200
Magnesium (mg/l)	-	50	30	-	100
Copper (mg/l)	1.3	1.0	0.05	1.5	1.5
Fluoride (mg/l)	4.0	1.5	0.6-1.2	1.5	1.5
Mercury (mg/l)	0.002	0.001	0.001	0.001	No relaxation
Cadmium (mg/l)	0.005	0.005	0.01	0.01	No relaxation
Selenium (mg/l)	0.05	0.01	-	-	No relaxation
Arsenic (mg/l)	0.05	0.05	0.05	0.05	No relaxation
Lead (mg/l)	-	0.05	0.10	0.05	No relaxation
Zinc (mg/l)	-	5.0	5.0	0.10	15.0
Chromium (mg/l)	0.1	-	0.05	-	No relaxation
E. coli (MPN/100 ml)	-	-	-	-	No relaxation

Fig. 3.4 International permissible water quality limits for various parameters (*Source* Kumar and Puri [2012])

The Rural Drinking Water and Sanitation Department (RDWSD) was created by the Karnataka government to provide clean drinking water and proper sanitation facilities in the rural areas through the Swachh Bharat Mission Gramin (SBMG) Karnataka and Jal Jeevan Mission (JJM) Karnataka programs(*RDWSD Karnataka—Official Website* 2021). The gram panchayat, taluk panchayat, zila panchayat, and state offices, along with RDWSD authorities, are tasked with conducting periodic check-ups of the groundwater quality biannually of the bore wells, tubes, and other sources available in the region. Nanda and Kumar (2015) in their analysis of groundwater quality in six village panchayats of Madhugiri taluk in Karnataka found that most of the water samples do not meet the standard water quality limits and, therefore, not suitable for consumption without proper treatment. Regardless of this concern, in many instances, it is observed that rural people consume water directly from the source with only to no treatment. A study on the Water Quality Index (WQI) of the Tumkuru taluk by Ramakrishnaiah et al. (2009) also revealed that the region's groundwater requires some treatment before consumption. The WQI of almost 99% of the total sample exceeded 100, which is the upper limit for drinking water. The water quality classification based on the WQI value is shown in Fig. 3.18. The water categorization assessment unit in the Tumkuru district as on March 2020 is shown in

Fig. 3.6 only 3 taluks out of 10 in Tumkur district have safe drinking water, indicating the need for urgent focus on the quality of water in the district (Fig. 3.5).

This study focuses on understanding the water consumption patterns in the two selected GPs of Madhugiri, the water monitoring and testing frequencies of the water quality by the authorities, and the use of government-provided RO-filtered water. Based on the findings and observations from the study, a framework is proposed to ensure regular water monitoring by the concerned authorities and information dissemination to the relevant stakeholders, especially consumers, through a common digitized report delivering information on the quality of drinking water. This will help

WQI Value	*Water quality*	*Percentage of water samples (Pre-monsoon)*
<50	*Execllent*	*00*
50-100	*Good water*	*1.5*
100-200	*Poor water*	*63.5*
200-300	*Vvery poor water*	*22*
>300	*Water unsuitablefor drinking*	*13.0*
Source: Ramakrishnaiah et al. (2009)		

Fig. 3.5 Water Quality classification based on WQI value (*Source* Ramakrishnaiah et al. [2009])

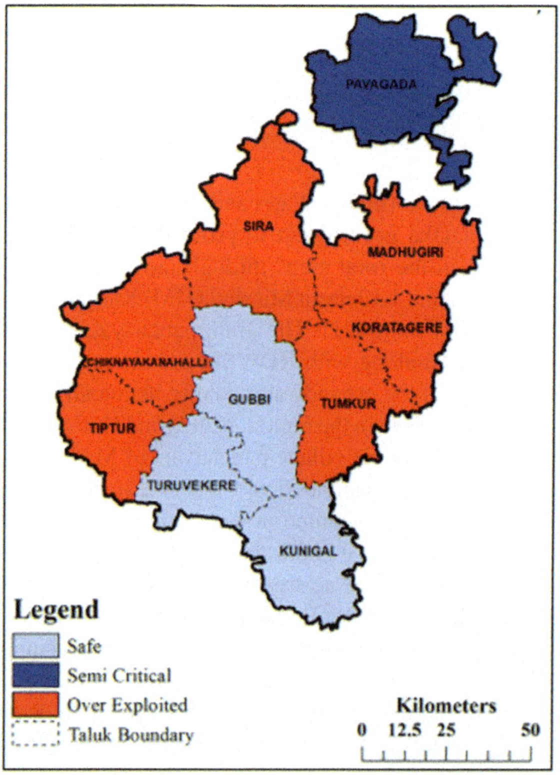

Fig. 3.6 Categorization of assessment Unit in Tumkuru district

consumers make a well-informed decision regarding the quality of water consumed and thus demand quality water from the local authorities. This exercise is likely to improve the performance of the water supply system in rural areas.

3.3 Methodology

This study is part of the ongoing action research program under Unnat Bharath Abhiyan. Field observations and unstructured interviews were conducted as part of the study to understand the grass-root-level condition of drinking water among the GPs. It provides a detailed description of the water sources, water consumption patterns, and the common issues faced by the people of Chinakavajra and DV Halli Gram Panchayats of Madhugiri Taluk. The water quality testing reports were collected from the RDWSD, Madhugiri, Tumkuru. The reports had samples from 14 villages of Chinakavajra Panchayat and 17 villages of DV Halli Panchayat. Unstructured interviews were conducted with the villagers and authorities to understand the drinking water issues and water testing frequency in the region.

The water test reports were collected from the RDWSD and the Panchayat Development Offices (PDOs). The water men (the person who is incharge of water distribution in the villages) were interviewed to understand the sample collection process. The RDSWD employees were also interviewed to understand the process of water sample collection, testing, and distribution of reports to the PDOs. The observations and interview results were compiled to depict the existing system and develop a framework for an efficient quality testing system to provide proper information to the stakeholders.

3.3.1 Study Area

Madhugiri taluk in the Tumkuru district is located in the eastern part of Karnataka (Fig. 3.7). This study focused on two selected GPs: DV Halli and Chinakavajra. Tumkuru shares its border with Ananthapur district of Andhra Pradesh. While Madhugiri taluk shares the border with Hindupur of Andhra Pradesh, Chinakavajra, and DV Halli is located at the southern boundary of Madhugiri. Chinakavajra GP consists of 14 villages, with a total population of 7222. DV Halli GP consists of 17 villages, with a total population of 6466. It is a drought-prone, rocky region with an average rainfall of 593.0 mm. The region usually receives rainfall from the southwest monsoon beginning in early June, which peaks during September. Groundwater is the primary source of drinking water, and rainfall is the only source of recharge in the region. According to Vittala (2017), the underground water table in this region is at a depth of about 1200 feet.

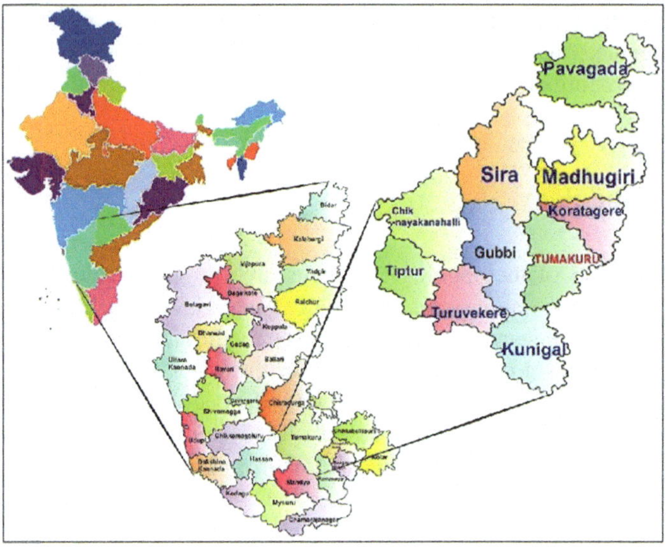

Fig. 3.7 Location of Madhugiri taluk in the district of Tumkuru shown in the map of India

3.3.2 Findings

The people of Chinakavajra and DV Halli of Madhugiri taluk mainly depend on borewells for drinking water. Generally, across GPs, a community water tap connected to the public borewell is installed at strategic locations, and people collect water from these taps. In Chinakavajra village, owing to the low availability of groundwater, local governments are supplying water through tankers to these villages. However, the water supplied through pipelines and water tankers is not filtered. Concerned by the issue that water supplied through pipelines could be contaminated, the local governments of each taluk have installed public water filter facilities in the villages, which can be accessed by people from three to four surrounding villages. Any villager can get filtered water for ₹5 for 20 liters. Presently, the two GPs have 42 borewells and 21 public reverse osmosis (RO) filter facilities; of which, only 11 ROs are properly functioning, whereas the others are closed owing to non-maintenance. However, most people in the two GPs of Madhugiri taluk are reluctant to utilize the RO filter water for drinking purposes and are found to be using the borewells and local government-supplied water for both drinking and general household purposes. Despite the harmful contaminants found in drinking water, people consume the same water directly from the sources, perhaps because of the absence of an efficient mechanism of sharing water quality–related information. However, only a few respondents reported that the drinking water they receive is contaminated and a negligible number stated that the water is salty; most respondents reported that the water is potable. Several other studies also mentioned reluctance for paying for drinking water from ROs, and a few others were hesitant in traveling long distances to collect water from the RO filters. The public RO water filter facilities are not periodically monitored or checked by the local government authorities. Several RO filters were found nonfunctional and in damaged conditions owing to poor maintenance. Hence, the water quality standards from the RO water filters cannot be completely trusted. In some villages, the Dairy Cooperative monitors the functioning and maintenance of the public RO water filters; however, in a majority of the other villages, the ROs are in poor condition and require maintenance. In some villages of both Chinakavajra and DV Halli Panchayats, where there is a low or depleting water level in the bore wells/tubes, the Panchayat authorities pay Rs.17,000 per month to purchase groundwater from farmers who have adequate water in the bore wells. The authorities supply water to other farmers for domestic and irrigation purposes. This is considered a good alternative for the people in Madhugiri region during summers and under acute water shortage situations. An examination of the expenditure profile of the two panchayats finds that drinking water does not constitute major funding in both regions—Chinakavajra (16%) and DV Halli (20%) (Figs. 3.8 and 3.9). The GPs spend most of the funding on sanitation work, which involves drainage cleaning, repairs, and toilet construction. However, there is less focus on allocating funds for alternative water resources, testing water quality, and making it potable.

Madaganahatti, a village in Chinakavajra panchayat, has seepage from a dump yard located within 2 km radius. The villagers of the panchayat have reported the issue owing to changes in the taste of water and odor. These villages have switched to RO water filters for domestic purposes. This suggests that the villagers' switch to or preference for RO is not based on the suggestions of the water testing authorities but on personal experiences of consuming polluted water. The RDWSD of Tumkuru in Karnataka must conduct biannual testing of groundwater quality, but currently, it is conducted only once a year. The authorities collect the sample from the sources and send it for testing to the laboratories. Subsequently, the reports are shared with

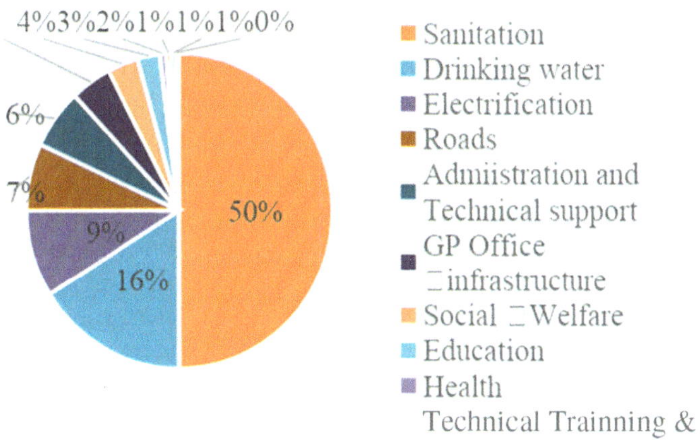

Fig. 3.8 Funds allocated (%) to different sectors in the past three years in Chinakavajra GP (*Source* Collected by the authors)

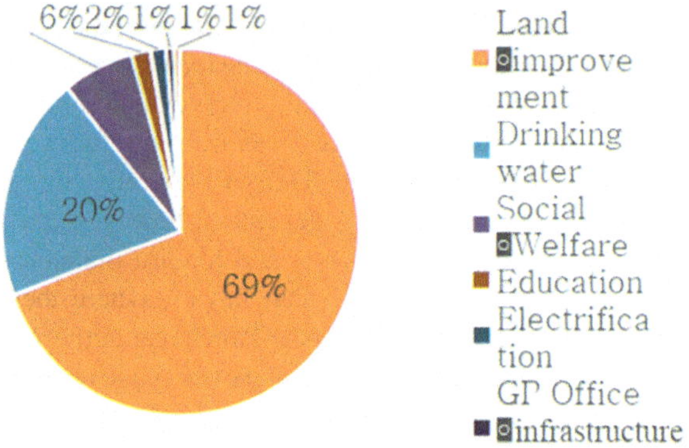

Fig. 3.9 Fund allocated (%) to different sectors in the past 3 years in DV Halli GP (*Source* Collected by the authors)

the panchayat development officers (PDOs), who in turn disseminate the information to the citizens through appropriate channels such as ward members and watermen. In the water testing reports, the potable sources are indicated as "PASS" and non-potable sources as "NOT PASS/NP." Owing to poor documentation, the department could provide water testing reports in Chinakavajra and DV Halli panchayats only for 2019, 2020, and 2021. The reports are shown in the Appendix (Figs. 3.14–3.18). In DV Halli, the reports showed four, six, and two non-potable sources in 2019, 2020, and 2021, respectively. In Chinakavajra, the reports showed that three, five, and two sources were unfit for consumption in 2019, 2020, and 2021, respectively. The sources from both GPs were marked as non-potable owing to high concentrations of fluoride found in the groundwater.

A 3-year analysis of the water reports indicates that the season during water testing has a significant effect on the test results. The drinking water samples collected during December 2019 and November 2021 showed similar results, while the samples collected in August 2020 showed a higher concentration of constituents in the groundwater (Table 3.1). Most of the physicochemical properties of the water samples showed higher values during the rainy season and lower values during winter. This indicates that the quality of groundwater in the region is influenced by seasonal variations (Vyas and Sawant 2008). The consolidated water testing report on the physicochemical properties of the water samples in the DV Halli panchayat for 2019, 2020, and 2021 are shown in Table 3.1 in the Appendix. There were several inconsistencies in documenting the "Village" name and "Location of source" over 3 years. Although each source in the village has a unique "Source ID," it is not documented systematically or is missing over the years.

Based on the field observations and the 2021 RDWSD water quality testing reports, a panchayat revealed certain mismatches which question the credibility of the reports released by the department. The testing department in the year 2021 has released two reports of the DV Halli panchayat where one report indicates two sources (R.N.Roppa & K.C.Roppa) as "PASS" (Figs. 3.10) and another report states the water from the same source as "NOT PASS" (Figs. 3.11). The discrepancy in the test reports questions the integrity of the department and the unsystematic dissemination

Table 3.1 Consolidated Water Quality Testing report of the DV Halli panchayat (*Note*: NA-Location of Source for the villages are not matching across the 3 years)

S.No	Village	Location of Source	Type of Source (BW—Surface well filter)	Name of Scheme (HP—MWS—PWS—MVS)	Fluoride (003)			Nitrate (006)			Total Dissolved Solids (TDS) mg/L (015)		
					as F (mg-L)			as NO_3 (mg-L)			500–2000		
					1.0–1.5			45–N.R					
					2019	2020	2021	2019	2020	2021	2019	2020	2021
1	D. V Halli	Near Avikatte road	BW	RO	NA	NA	0.75	NA	NA	8	NA	NA	64
2	D. V Halli	Near Roadside	BW	MWS	NA	NA	0.92	NA	NA	14	NA	NA	262
3	D. V Halli	Near Padagappa House	BW	MWS	NA	NA	1.42	NA	NA	16	NA	NA	410
4	D. V Halli	Near land	BW	OHT	NA	NA	1.1	NA	NA	14	NA	NA	393
5	Achenahalli	in the village	BW	OHT	1.02	1.75	1.3	22	10	10	675	435	272
6	Achenahalli	Near Guddaranganatha swamy	BW	MWS	1.75	1.75	1.1	31	19	13	729	748	188
7	Gurummana katte	Near kere	BW	MWS	1.23	NA	1.2	22	NA	12	752	NA	196
8	Gurummana katte	Near Narasimiah House	BW	MWS	1.24	NA	1.21	11	NA	8	707	NA	266
9	Thayegondanahalli	Near Hosaband	BW	MWS	NA	NA	1.3	NA	NA	20	NA	NA	298
10	Thayegondanahalli	Near Aralekatte	BW	MWS	NA	NA	1.2	NA	NA	18	NA	NA	340
11	R. N. Roppa	Near Maramma temple	BW	MWS	0.65	NA	1.57	26	NA	22	428	NA	481

(continued)

Table 3.1 (continued)

S.No	Village	Location of Source	Type of Source (BW—Surface well filter)	Name of Scheme (HP—MWS—PWS—MVS)	Fluoride (003)			Nitrate (006)			Total Dissolved Solids (TDS) mg/L (015)		
					as F (mg-L)			as NO_3 (mg-L)			500–2000		
					1.0–1.5			45–N.R					
					2019	2020	2021	2019	2020	2021	2019	2020	2021
12	K. C. Roppa	Near Maramma temple	BW	OHT	1.23	1.95	1.59	38	14	25	860	878	633
13	Belladamadugu	Near Jayaramma house	BW	MWS	1.52	NA	1.22	25	NA	18	828	NA	230
14	Somalara	In the village	BW	OHT	NA	NA	1.3	NA	NA	11	NA	NA	210
15	Somalara	Near Kote Rangappa temple	BW	MWS	NA	NA	1.11	NA	NA	10	NA	NA	163
16	Gundlahalli	Near Gowda house	BW	MWS	1.35	NA	1.2	26	NA	12	342	NA	252
17	Gundlahalli	Near kere	BW	MWS	1.75	NA	0.87	24	NA	10	527	NA	180
18	Gundlahalli	Near Kallapa land	BW	OHT	1.11	NA	1.4	26	NA	19	680	NA	295

Zilla Panchayath Tumakuru
Taluk- Madhugiri

Water Quality Testing Report
Based on IS-10500 . 2012

R.D.W.S & Sanitation Division , Tumakuru

Sl.No	GP	village	Location of source	Type of Source (BW-Surface well filter)	Name of Scheme (HP-MWS-PWS-MVS	Date of collection and testing	pH Value (013)	Turbidity (001) (NTU)	Total Dissolved Solids (TDS)	Nitrate (006) as NO3 (mg/l)	Flouride (003) as(F) (mg/l)	Chloride (005) (as Cl) (mg/l)	Total Alkalinity (018) as CaCO3 (mg/l)	Total Hardness (016) as CaCO3 (mg/l)	Sulphate (007) as SO4 (mg/l)	Total Iron (004) as Fe (mg/l)	Calcium (as Ca) (mg/l)	Remarks (P-NP-PAAS)
							6.5-8.5	(1-5)	500-2000	45 N R	1.0-1.5	[illegible]	[illegible]	[illegible]	[illegible]	[illegible]	75-200	
1	D V Halli	D V Halli	Near land	BW	MWS	10.11.21	7.5	1.00	393	14	1.1	155	210	204	103	0.06	82	PAAS
2	D V Halli	Achenahalli		BW	RO	10.11.21	7.9	1.00	272	10	1.3	142	120	110	98	0.09	64	PAAS
3	D V Halli	D V Halli	Near Avikatte road	BW	PWS	10.11.21	7.3	1.00	64	8	0.75	35	22	30	12	0.03	12	PAAS
4	D V Halli	Gurummana katte	Near kere	BW	PWS	10.11.21	7.5	1.00	196	12	1.2	102	112	106	76	0.07	42	PAAS
5	D V Halli	Thayegondanahalli	Near Hosabande	BW	PWS	10.11.21	7.4	1.00	298	20	1.3	104	150	140	60	0.09	56	PAAS
6	D V Halli	Thayegondanahalli	Near Aralekatte	BW	PWS	10.11.21	7.2	1.00	340	18	1.2	123	180	160	99	0.08	64	PAAS
7	D V Halli	B N ROPPA	Near Maramma temple	BW	PWS	10.11.21	7.5	2.00	481	22	1.57	210	248	240	124	0.06	96	PAAS
8	D V Halli	D V Halli	Near road side	BW	RO	10.11.21	7.1	1.00	262	14	0.92	180	210	200	110	0.07	80	PAAS
9	D V Halli	Achenahalli	Near Guddaranganatha …	BW	PWS	10.11.21	7.3	1.00	188	13	1.1	108	90	80	60	0.04	32	PAAS
10	D V Halli	Gurummana katte	Near Narasimhaiah house	BW	RO	10.11.21	7.9	1.00	266	8	1.21	110	140	128	77	0.06	50	PAAS
11	D V Halli	D V Halli	Near padagappa house	BW	PWS	10.11.21	7.5	1.00	410	16	1.42	135	220	210	90	0.09	84	PAAS
12	D V Halli	Belladamadugu	Near Jayaramanna hous…	BW	PWS	10.11.21	7.8	1.00	230	18	1.22	130	128	118	100	0.08	46	PAAS
13	D V Halli	Somalara	Near kote Rangappa tem…	BW	PWS	10.11.21	7.4	1.00	163	10	1.11	98	100	95	75	0.10	18	PAAS
14	D V Halli	K C ROPPA	Near Maramma temple	BW	MWS	10.11.21	7.3	2.00	633	25	1.59	288	388	360	182	0.12	144	PAAS
15	D V Halli	Gundlahalli	Near Gowda house	SW	PWS	10.11.21	7.6	1.00	252	12	1.2	169	130	119	94	0.06	48	PAAS
16	D V Halli	Somalara	in the village	BW	PWS	10.11.21	7.2	1.00	210	11	1.3	90	108	102	86	0.09	40	PAAS
17	D V Halli	Gundlahalli	Near kere	BW	RO	10-11-21	7.5	1.00	180	10	0.87	109	132	128	82	0.07	48	PAAS
18	D V Halli	Gundlahalli	Near kallappa land	BW	MWS	10.11.21	7.2	1.00	295	19	1.4	150	149	139	105	0.06	56	PAAS

RDWSD
Water Quality Lab

Fig. 3.10 Water Quality Testing Report-1 of DV Halli Panchayat (2021) *Source* Collected by the authors)

of information to the stakeholders. Moreover, there is lack of coordination and cooperation between the Panchayat offices and the RDWSD personnel, where neither the tests are conducted systematically, nor the reports promptly shared with the public. Besides, there is the issue of frequent changes in the water distribution men owing to local politics. These water men are responsible for water distribution and assisting water testing agencies in finding the right source according to the Source ID. Thus, the water-distributing person/s is not completely aware about the region, the number of water sources in each village, and the source code. This discrepancy perhaps is one of the reasons leading to incorrect water quality testing reports. From these above observations, it is evident that the water quality testing reports from RDWSD cannot be regarded as a reliable source of information to assess the quality of groundwater from a particular source.

3.3.3 Current Process of Sample Collection, Testing, and Information Dissemination

The water testing centers under the RDWSD, Karnataka, are authorized to collect water samples from the source(s) across the districts of the state. The water sample collectors seek help from a villager responsible for the water distribution in the village because of his familiarity with the sources. The water collected from the source is taken to the testing centers of each district. The water quality testing reports are then shared with the Panchayat Development Officers (PDOs) for information dissemination to the consumers. Delivery of the water test reports to the PDO's

RWS&SD Water Quality Parameters Tested In Sub-Divisional Lab - Madhugiri , Tumkur District - Water Sample Collection And Analysis Results. Tq -Madhugiri

Panchayath :D.V.HALLI

Sl.No	village	Location of source	Type of Source (BW-Surface well filter)	Name of Scheme (HP-MWS-PWS-MVS)	Date of collection and testing	pH Value (013) 6.5-8.5	Turbidity (001) (NTU) (1-5)	Total Dissolved Solids (TDS) mg/L (015) 500-2000	Nitrate (006) as NO3 (mg/L) 45-N.R.	Flouride (013) as F (mg/L) 1.0-1.5	Chloride (005) (as Cl) (mg/L) 250-1000	Total Alkalinity (018) as CaCo3 (mg/L) 200-600	Total Hardness (016) as CaCo3 (mg/L) 200-600	Sulphate (007) as SO4 (mg/L) 200-400	Total Iron (004) (as Fe) (mg/L) 1.0-N.R	Calcium (as Ca) (mg/L) 75-200	Remarks (P-NP-PAAS)
1	D V Halli	Near land	BW	OHT	10-11-21	7.50	1.00	393	14	1.10	155	210	204	103	0.06	82	PAAS
2	Achenahalli	In the village	BW	OHT	10-11-21	7.90	1.00	272	10	1.30	142	120	110	98	0.08	44	PAAS
3	D V Halli	Near Avikatte road	BW	RO	10-11-21	7.30	1.00	64	8	0.75	35	22	30	12	0.03	12	PAAS
4	Gurummana katte	Near kere	BW	MWS	10-11-21	7.50	1.00	196	12	1.20	102	112	106	76	0.07	42	PAAS
5	Thayegondanahalli	Near Hosaband	BW	MWS	10-11-21	7.40	1.00	298	20	1.30	104	150	140	80	0.09	56	PAAS
6	Thayegondanahalli	Near Aralekatte	BW	MWS	10-11-21	7.20	1.00	340	18	1.20	123	180	160	99	0.08	64	PAAS
7	R.N.Roppa	Near Maramma temple	BW	MWS	10-11-21	7.50	1.00	481	22	1.57	210	248	240	124	0.06	96	NP
8	D.V Halli	Near Road side	BW	MWS	10-11-21	7.10	1.00	262	14	0.92	180	210	200	110	0.07	80	PAAS
9	Achenahalli	Near Goddaranganathswamy	BW	MWS	10-11-21	7.30	1.00	188	13	1.10	108	90	80	60	0.04	32	PAAS
10	Gurummana katte	Near Narasimiah House	BW	MWS	10-11-21	7.90	1.00	266	8	1.21	110	140	128	77	0.06	50	PAAS
11	D.V Halli	Near Padagappa house	BW	MWS	10-11-21	7.50	1.00	410	16	1.42	135	220	210	90	0.09	84	PAAS
12	Belladamadugu	Near Jayaramana house	BW	MWS	10-11-21	7.80	1.00	230	18	1.22	130	128	118	100	0.08	46	PAAS
13	Somalara	Near Kote Rangappa temple	BW	MWS	10-11-21	7.40	1.00	163	10	1.11	98	100	95	75	0.10	38	PAAS
14	K.C.Roppa	Near Maramma temple	BW	OHT	10-11-21	7.30	1.00	633	25	1.59	288	388	360	182	0.12	144	NP
15	Gandlahalli	Near Gowda house	BW	MWS	10-11-21	7.60	1.00	252	12	1.20	169	130	119	94	0.06	48	PAAS
16	Somalara	In the village	BW	OHT	10-11-21	7.20	1.00	210	11	1.30	90	108	102	86	0.09	40	PAAS
17	Gandlahalli	Near kere	BW	MWS	10-11-21	7.50	1.00	180	10	0.87	109	132	128	82	0.07	48	PAAS
18	Gandlahalli	Near Kallappa land	BW	OHT	10-11-21	7.20	1.00	295	19	1.40	150	149	139	105	0.06	56	PAAS

Chemist
RDWSD
Water Quality Lab
Madugiri

Fig. 3.11 Water Quality Testing Report-2 of DV Halli Panchayat (2021) (*Source* Collected by the authors)

office is a challenge owing to noncooperation between the authorities regarding the collection and delivery of the report. The PDOs communicate the test results to the water distribution authorities and/or the elected panchayat members during the general body meeting held quarterly.

Finally, information on portability is shared by the concerned officials with the consumers of the district. This conventional method of information dissemination is a long process and includes the involvement of several parties, which increases the risk of human error and data tampering. To address these challenges, we propose a framework to ensure smooth and efficient dissemination of information regarding water quality to the citizens/consumers. Figure 3.12 shows the conventional method followed to disseminate the information regarding water quality.

3.4 Suggestions for Improvement

In the proposed framework, the water sample collection from the sources and the testing follows a similar pattern to the conventional method. In the next stage of the information dissemination channel, the water quality testing reports can be uploaded to the RDWSD's website. This simple method of information dissemination through

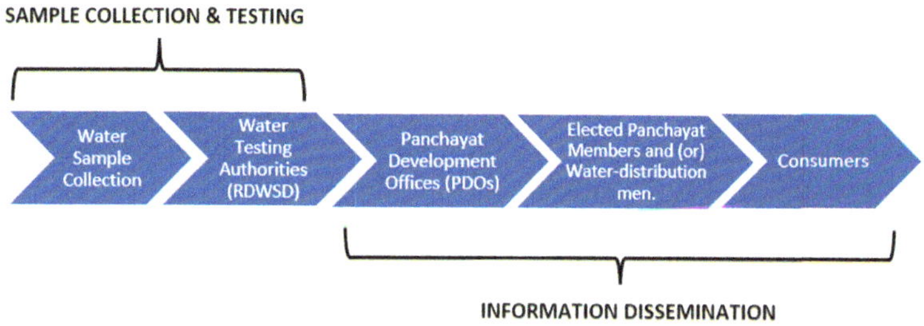

Fig. 3.12 Conventional method of information dissemination

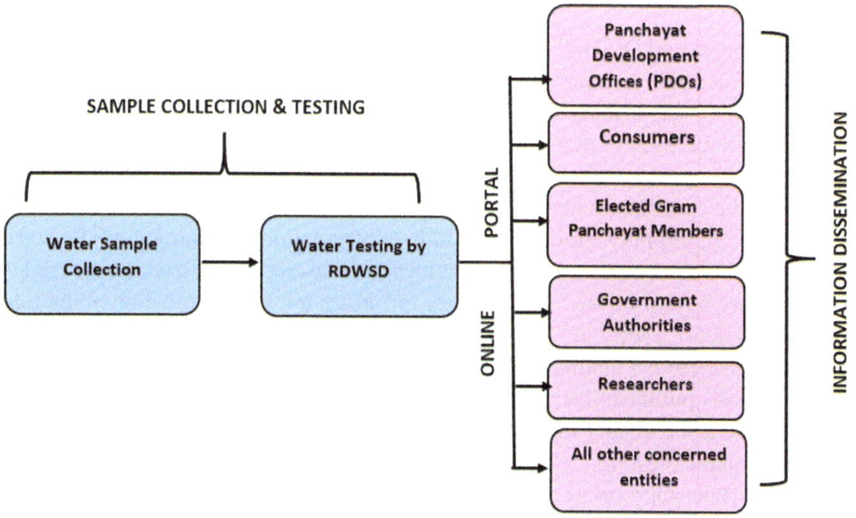

Fig. 3.13 Proposed method of information dissemination by the authors

a digitized report on the official website addresses the challenges faced by the conventional method. It also helps different information seekers such as the PDO's office, consumers, researchers, government authorities, elected members, and others to access the test reports as required. The proposed framework will also help in resolving the disagreement between the PDO and the RDWSD about sharing and delivering test reports. It will also ensure that the water testing authorities test periodically and regularly share information with the relevant stakeholders. Figure 3.13 shows the proposed framework for the dissemination of information to consumers.

Digitization of the information dissemination process simplifies the long conventional method and streamlines the functioning of the entire system. Increasing the frequency of water testing by the department, proper documentation of the Source IDs in the testing reports, and periodically updating the reports in the portal will provide more information to assess the water quality in the region. This will also create a platform for increasing awareness among the public on the issues of drinking water and focuses on the discussion during the GP meetings of the taluk.

3.5 Conclusion

In Chinakavajra and DV Halli Panchayats of Madhugiri taluk in Karnataka, access to safe and clean drinking water is constrained owing to the poor quality of groundwater in the region. Based on the unstructured interviews and test reports, we have identified that several sources have non-potable water owing to high concentrations of fluoride content in the groundwater. Therefore, it needs to be tested frequently and treated properly. The coordination between the water testing authorities and the PDOs was also fragile. Importantly, the current system of information dissemination followed in the region is unsystematic and there is evidence of discrepancy. To address these key concerns, we have proposed a framework for information dissemination to the relevant stakeholders, especially the users, through a common digitized report providing information about the quality of drinking water. The principle behind making water quality reports public and accessible to consumers is that consumers have the right to know about the quality of their drinking water and to increase the credibility of the water quality testing reports issued by the department. Informed and involved citizens can take appropriate decisions and ensure the proper functioning of the water supply system.

Appendix

See Figs. 3.14, 3.15, 3.16, 3.17 and 3.18; Tables 3.2, 3.3, 3.4, 3.5 and 3.6

Zilla Panchayath Tumakuru
Taluk: MADHUGIRI
Panchayath: CHINAKAVAJRA

Water Quality Testing Report
Based on IS-10500 : 2012

R.D.W.S & Sanitation Division, Tumakuru

Sl. No	village	Location of source	Type of Source (BW/Surface well/filter)	Name of Scheme (HP/MWS/PWS/MVS)	Date of collection and testing	pH Value	Turbidity (NTU)	Total Dissolved Solids (TDS) mg/L	Nitrate as NO3 (mg/L)	Flouride as (F) (mg/L)	Chloride as Cl (mg/L)	Total Alkalinity as CaCO3 (mg/L)	Total Hardness as CaCO3 (mg/L)	Sulphate as SO4 (mg/L)	Total Iron as Fe (mg/L)	Calcium as Ca (mg/L)	Remarks (P-NP-PAAS)
						6.5-8.5	(1-5)	500-2000	45-N.A.	1.0-1.5	250-1000	200-600	200-600	200-400	1.0-N.A.	75-200	
1	Gudiroppa	near Hariharashwara temple	BW	MWS	13/12/2019	8.09	0.05	252	12	1.21	78	210	180	40	0.19	72	PAAS
2	Keregalapalya	near Kere	BW	MWS	13/12/2019	7.84	1	372	30	1.42	92	286	248	42	0.16	99	PAAS
3	Keregalapalya	near Kere Dhabba	BW	MWS	13/12/2019	7.54	0.05	343	20	1.28	105	226	228	40	0.19	91	PAAS
4	Chinakavajra	near Forest boundary	BW	MWS	13/12/2019	7.58	0.05	300	14	1.26	96	105	128	42	0.13	51	PAAS
5	Keregalapalya	near Kavadigara koppa	BW	MWS	13/12/2019	7.44	1	383	12	1.31	115	260	240	70	0.14	96	PAAS
6	Chinakavajra	Main road	BW	MWS	13/12/2019	7.25	1	625	26	1.51	220	402	384	90	0.18	154	NP
7	Kataganahatti	Main road	BW	MWS	13/12/2019	7.29	0.05	371	21	1.39	181	180	164	68	0.11	66	PAAS
8	Kataganahatti	near Kataganahatti gate	BW	MWS	13/12/2019	7.35	1	452	18	1.28	126	228	268	60	0.11	107	PAAS
9	Kambaiahnapalya	in the village	BW	MWS	13/12/2019	7.25	0.05	463	19	1.46	128	264	282	68	0.12	113	PAAS
10	Hunasemaradahatti	near Eranna house	BW	MWS	13/12/2019	7.04	0.05	646	23	1.61	302	380	402	120	0.14	161	NP
11	Maribeelu	right side of road	BW	MWS	13/12/2019	7.34	0.05	360	22	1.31	105	228	210	64	0.16	84	PAAS
12	Maribeelu	left side of road	BW	MWS	13/12/2019	7.24	1	363	20	1.28	98	246	240	70	0.12	96	PAAS
13	Gudiroppa	in the village	BW	MWS	13/12/2019	7.45	1	252	18	1.4	92	164	184	68	0.08	74	PAAS
14	Madaganahatti	near Halla	BW	MWS	13/12/2019	7.34	1	854	19	1.71	284	468	440	160	0.26	176	NP

Fig. 3.14 Water quality testing report of Chinakavajra Panchayat (2019) (*Source* Collected by the authors)

Zilla Panchayath Tumakuru
Taluk: MADHUGIRI
Panchayath: D.V.Halli

Water Quality Testing Report
Based on IS-10500 : 2012

R.D.W.S & Sanitation Division, Tumakuru

Sl. No	village	Location of source	Type of Source (BW/Surface well/filter)	Name of Scheme (HP/MWS/PWS/MVS)	Date of collection and testing	pH Value	Turbidity (NTU)	Total Dissolved Solids (TDS) mg/L	Nitrate as NO3 (mg/L)	Flouride as (F) (mg/L)	Chloride as Cl (mg/L)	Total Alkalinity as CaCO3 (mg/L)	Total Hardness as CaCO3 (mg/L)	Sulphate as SO4 (mg/L)	Total Iron as Fe (mg/L)	Calcium as Ca (mg/L)	Remarks (P-NP-PAAS)
						6.5-8.5	(1-5)	500-2000	45-N.A.	1.0-1.5	250-1000	200-600	200-600	200-400	1.0-N.A.	75-200	
1	D V Halli	near Shivanna house	BW	MWS	13/12/2019	7.06	1	356	20	0.95	118	230	380	32	0.12	140	PAAS
2	Kuppachari roppa	near Maramma temple	BW	MWS	13/12/2019	6.8	1	860	38	1.58	225	520	460	95	0.14	180	NP
3	Gundlahalli	near Kere	BW	MWS	13/12/2019	7.16	0.05	527	24	1.75	140	488	330	58	0.16	135	NP
4	Achenahalli	near Ranganathaswamy tema	BW	MWS	13/12/2019	7.28	0.05	729	31	1.75	180	490	440	81	0.11	160	NP
5	Thangoti	near Shivanna land	BW	MWS	13/12/2019	7.42	1	590	30	1.25	155	420	335	65	0.12	130	PAAS
6	Kambadhahalli	near Road	BW	MWS	13/12/2019	7.35	1	405	24	0.89	130	330	280	45	0.13	60	PAAS
7	Gundlahalli	near Gowda's house	BW	MWS	13/12/2019	7.26	1	342	26	1.35	70	360	250	38	0.14	50	PAAS
8	Achenahalli	near Kambadhahalli road	BW	MWS	13/12/2019	7.27	1	675	22	1.02	110	630	455	65	0.16	130	PAAS
9	Gurrammanakatte	in the kere	BW	MWS	13/12/2019	7.24	1	752	22	1.23	215	440	380	85	0.12	195	PAAS
10	R N Roppa	near Thimmadhasappa land	BW	MWS	13/12/2019	7.32	0.05	428	26	0.65	90	430	310	48	0.14	65	PAAS
11	Gundlahalli	near Kallappa land	BW	MWS	13/12/2019	7.17	1	680	26	1.11	160	520	480	75	0.16	135	PAAS
12	Kuppachari roppa	near road	BW	MWS	13/12/2019	7.09	1	558	22	0.95	115	450	360	62	0.14	70	PAAS
13	Gurrammanakatte	near Narasimhaiah house	BW	MWS	13/12/2019	7.07	1	707	11	1.24	190	520	450	78	0.12	130	PAAS
14	Hosahalli	near Manjanna house	BW	MWS	13/12/2019	7.06	1	293	1	1.15	90	240	165	32	0.14	35	PAAS
15	Somlara	near Road	BW	MWS	13/12/2019	7.04	0.05	698	6	1.45	215	460	390	77	0.16	135	PAAS
16	Thayigondanahalli	near Poojar land	BW	MWS	13/12/2019	7.12	1	572	40	0.85	165	520	30	64	0.12	85	PAAS
17	Belladhamadagu gate	near Muddarangappa house	BW	MWS	13/12/2019	7.24	1	828	25	1.52	230	550	530	92	0.14	90	NP

Fig. 3.15 Water quality testing report of DV Halli Panchayat (2019) (*Source* Collected by the authors)

RWS&SD Water Quality Parameters Tested In Sub-Divisional Lab - Madhugiri , Tumkur District - Water Sample Collection And Analysis Results. Tq -Madhugiri

Panchayath :Chinakavajra

Sl.No	village	Location of source	Source ID	Type of Source (BW-Surface well filter)	Name of Scheme (HP-MWS-PWS-MVS)	Date of collection and testing	pH Value (013) 6.5-8.5	Turbidity (NTU) (1-5)	Total Dissolved Solids (TDS) mg/L (015) 500-2000	Nitrate (006) as NO3 (mg/L) 45-N.R.	Fluoride (001) as F (mg/L) 1.0-1.5	Chloride (005) (as Cl) (mg/L) 250-1000	Total Alkalinity (018) as CaCO3 (mg/L) 200-600	Total Hardness (016) as CaCO3 (mg/L) 200-600	Sulphate (007) as SO4 (mg/L) 200-400	Total Iron (004) (as Fe) (mg/L) 1.0-N.R.	Calcium (as Ca) (mg/L) 75-200	Remarks (P-NP-PAAS)
1	Keregala palya	near swanenahalli kere	250056231	BW	OHT	22-08-20	6.96	2.00	660	31	1.97	168	410	316	107	0.11	126	NP
2	Keregala palya	near karadappa temple	5164835	BW	OHT	22-08-20	6.80	2.00	450	28	1.60	140	486	218	96	0.15	102	NP
3	Keregala palya	kawadirapa bore	250056238	BW	MWS	22-08-20	6.90	2.00	810	35	1.67	220	540	324	118	0.09	154	NP
4	Keregala palya	near anganavadi janthakeri	271646	BW	MWS	22-08-20	7.23	2.00	1020	12	1.72	312	574	408	175	0.14	149	NP
5	Keregala palya	near tumkur road	278626	BW	MWS	22-08-20	7.22	2.00	890	14	1.46	266	452	574	122	0.08	198	PAAS
6	HunseMaradhaHalli	near HMH main road	250236771	BW	MWS	22-08-20	6.85	2.00	790	9	1.45	244	324	374	110	0.12	150	PAAS
7	HunseMaradhaHalli	near jhankal road	250115429	BW	MWS	22-08-20	7.12	2.00	648	13	1.34	284	394	376	77	0.11	164	PAAS
8	Katagana Hatti	near kataganahatti main road	5541002	BW	MWS	22-08-20	7.10	2.00	710	32	1.06	210	458	364	83	0.14	162	PAAS
9	Katagana Hatti	near kataganahatti main road	250053865	BW	MWS	22-08-20	7.12	2.00	390	22	1.21	180	202	292	78	0.12	105	PAAS
10	Katagana Hatti	near tumkur road	5541030	BW	MWS	22-08-20	7.37	2.00	410	13	1.34	246	222	264	86	0.11	110	PAAS
11	Kambayana Palya	near temple	250056230	BW	MWS	22-08-20	7.46	2.00	540	16	1.30	208	242	334	118	0.14	100	PAAS
12	Chinakavjra J C	near gadiropa road	250056228	BW	MWS	22-08-20	8.25	2.00	390	21	1.21	234	222	356	85	0.11	115	PAAS
13	Chinakavjra J C	In the village	250008971	BW	MWS	22-08-20	8.07	2.00	580	32	1.24	182	268	374	117	0.12	118	PAAS
14	Chinakavjra	near charanavanadhurga road	274689	BW	OHT	22-08-20	7.63	2.00	456	26	0.96	168	220	234	120	0.16	102	PAAS
14	Madaganahatti	In the village		BW	MWS	24-08-21	7.66	2.00	658	20	1.53	210	324	328	114	0.12	130	NP

Rural Water Supply & Sanitation Sub-Division Laboratory

Fig. 3.16 Water quality testing report of Chinakavajra Panchayat (2020) (*Source* Collected by the authors)

RWS&SD Water Quality Parameters Tested In Sub-Divisional Lab - Madhugiri , Tumkur District - Water Sample Collection And Analysis Results. Tq -Madhugiri

Panchayath :D.V.Halli

Sl.No	village	Location of source	Source ID	Type of Source (BW-Surface well filter)	Name of Scheme (HP-MWS-PWS-MVS)	Date of collection and testing	pH Value (013) 6.5-8.5	Turbidity (NTU) (1-5)	Total Dissolved Solids (TDS) mg/L (015) 500-2000	Nitrate (006) as NO3 (mg/L) 45-N.R.	Fluoride (001) as F (mg/L) 1.0-1.5	Chloride (005) (as Cl) (mg/L) 250-1000	Total Alkalinity (018) as CaCO3 (mg/L) 200-600	Total Hardness (016) as CaCO3 (mg/L) 200-600	Sulphate (007) as SO4 (mg/L) 200-400	Total Iron (004) (as Fe) (mg/L) 1.0-N.R.	Calcium (as Ca) (mg/L) 75-200	Remarks (P-NP-PAAS)
1	D.V.Halli	Near Belldamadugu road	250009261	BW	MWS	28-08-20	6.80	1.00	891	33	1.05	268	355	446	101	0.16	178	pass
2	D.V.Halli	Near Sira road	259236776	BW	OHT	28-08-20	6.90	1.00	1019	28	1.45	304	406	508	94	0.11	200	pass
3	D.V.Halli	Near Bikkanahope road	250056260	BW	MWS	28-08-20	6.60	1.00	1006	18	1.21	218	404	362	170	0.09	145	pass
4	D.V.Halli	Near Timblapura road	250179395	BW	OHT	28-08-20	6.80	0.05	723	24	1.24	290	292	562	120	0.06	725	pass
5	D.V.Halli	Near Thangatti road	250045395	BW	OHT	28-08-20	7.00	1.00	966	28	1.15	132	388	590	74	0.12	236	pass
6	AchenaHalli	Near Godadha Ranganatha S temple	250236843	BW	MWS	28-08-20	6.20	1.00	748	19	1.75	292	472	374	80	0.16	150	NP
7	AchenaHalli	Near KambadaHalli road	250056288	BW	OHT	28-08-20	6.80	0.05	435	10	1.75	210	300	500	75	0.13	200	NP
8	D.V.Halli	Near KambadaHalli road	250022844	BW	OHT	28-08-20	7.10	1.00	840	12	1.25	280	406	600	80	0.07	374	pass
9	GundlaHalli	Near Siddapura kere	250027804	BW	OHT	28-08-20	6.70	1.00	699	17	0.91	262	278	462	115	0.03	185	pass
10	GundlaHalli	Near Somalar road	5542334	BW	MWS	28-08-20	6.90	0.05	782	24	1.55	172	336	420	115	0.18	166	NP
11	GundlaHalli	Near Somalar road	280363	BW	MWS	28-08-20	6.90	1.00	1125	22	1.16	160	280	350	115	0.21	149	pass
12	GundlaHalli	Near Somalar road	250056261	BW	MWS	28-08-20	6.60	1.00	619	35	0.91	206	248	466	120	0.12	186	pass
13	GundlaHalli	Near Bettada road	250009995	BW	MWS	28-08-20	6.60	1.00	430	21	1.21	138	414	436	95	0.16	174	pass
14	GundlaHalli	Near Gutammana katte kere	250219943	BW	MWS	28-08-20	6.70	1.00	397	18	1.00	204	338	340	95	0.14	124	pass
15	GundlaHalli	Near JamsaHalli katte	3567571	BW	MWS	28-08-20	7.10	1.00	571	26	1.55	182	298	518	97	0.08	200	NP
16	GundlaHalli	Near Anganavadi kendra	250008103	BW	MWS	28-08-20	6.80	1.00	1156	24	1.75	135	440	422	96	0.06	169	NP
17	GundlaHalli	Near Kunachitoppa road	281714	BW	MWS	28-08-20	6.90	1.00	878	14	1.95	173	274	288	96	0.07	113	NP
18	GundlaHalli	Near Havekatte road	284364	BW	MWS	28-08-20	7.10	1.00	528	12	1.13	198	486	224	91	0.14	110	pass
19	GundlaHalli	Near Havekatte road	5284213	BW	MWS	28-08-20	7.40	1.00	684	10	1.12	205	242	402	80	0.11	161	pass

Rural Water Supply & Sanitation Sub-Division Laboratory

Fig. 3.17 Water quality testing report of DV Halli Panchayat (2020) (*Source* Collected by the authors)

RWS&SD Water Quality Parameters Tested In Sub-Divisional Lab - Madhugiri , Tumkur District - Water Sample Collection And Analysis Results Tq -Madhugiri

Panchayath :Chinakavajra

Sl.No	village	Location of source	Type of Source (BW-Surface well filter)	Name of Scheme (HP-MWS-PWS-MVS)	Date of collection and testing	pH Value (013)	Turbidity (001) (NTU)	Total Dissolved Solids (TDS) mg/L (015)	Nitrate (006) as NO3 (mg-L)	Fluoride (013) as(F) (mg-L)	Chloride (005) (as Cl) (mg-L)	Total Alkalinity (018) as CaCo3 (mg-L)	Total Hardness (016) as CaCo3 (mg-L)	Sulphate (007) as SO4 (mg-L)	Total Iron (004) (as Fe) (mg-L)	Calcium (as Ca) (mg-L)	Remarks (P-NP-PAAS)
						6.5-8.5	(1-5)	500-2000	45-N.R	1.0-1.5	250-1000	200-600	200-600	200-400	1.0-N.R	75-200	
1	Keregalapalya	Near kere Bande	BW	MWS	02-11-21	7.80	1.00	310	12	1.40	120	220	228	60	0.09	91	PAAS
2	Katagaranahalli	Near pouil	BW	OHT	02-11-21	7.50	1.00	264	14	1.20	88	115	120	64	0.08	48	PAAS
3	Kammanakote	Near pwd road	BW	MWS	02-11-21	6.90	1.00	260	10	1.30	105	120	110	40	0.07	44	PAAS
4	Marebelu	Near road wright side	BW	OHT	02-11-21	7.20	1.00	215	10	1.20	100	140	123	50	0.12	53	PAAS
5	Chinakavagra	Near Main road side	BW	OHT	02-11-21	6.90	1.00	393	13	1.40	130	228	224	80	0.11	90	PAAS
6	Gudiroappa	Near SC Colony	BW	MWS	02-11-21	7.80	1.00	250	12	1.10	120	164	168	88	0.07	67	PAAS
7	Hanasemaradha hatti	Near Janakal road	BW	OHT	02-11-21	7.70	1.00	298	10	1.10	110	140	120	40	0.08	48	PAAS
8	Kavadegara palya	In the village	BW	OHT	02-11-21	8.10	1.00	294	13	1.10	90	146	132	64	0.09	53	PAAS
9	Madagana hatti	Near Halla	BW	OHT	02-11-21	6.90	1.00	475	14	1.60	220	280	268	78	0.21	107	NP
10	Chinakavagra	Near Channarayan durga road	BW	MWS	02-11-21	6.80	1.00	110	9	0.98	95	58	60	24	0.05	24	PAAS
11	Gudiroappa	In the village	BW	MWS	02-11-21	7.40	1.00	208	10	1.00	80	105	110	40	0.06	44	PAAS
12	Marebelu	Near road side	BW	MWS	02-11-21	7.20	1.00	214	10	1.10	85	128	138	58	0.04	55	PAAS
13	Keregalapalya	Near Janatha colony	BW	MWS	02-11-21	7.30	1.00	200	9	1.10	88	105	100	36	0.04	40	PAAS
14	Kambayanapalya	In the village	BW	MWS	02-11-21	7.50	1.00	375	15	1.70	210	184	160	60	0.05	64	NP

Fig. 3.18 Water quality testing report of Chinakavajra Panchayat (2021) (*Source* Collected by the authors)

Table 3.2 Inorganic contaminants found in water

Contaminant	Sources to groundwater	Potential health and other effects
Aluminum	Occurs naturally in some rocks and drainage from mines	Can precipitate out of water after treatment, causing increased turbidity or discolored water
Antimony	Enters the environment from natural weathering, industrial production, municipal waste disposal, and manufacturing of flame retardants, ceramics, glass, batteries, fireworks, and explosives	Decreases longevity, alters blood levels of glucose and cholesterol in laboratory animals exposed to high levels over their lifetime
Arsenic	Enters the environment from natural processes, industrial activities, pesticides, and industrial waste, smelting of copper, lead, and zinc ore	Causes acute and chronic toxicity, liver and kidney damage; decreases blood hemoglobin. A carcinogen
Barium	Occurs naturally in some limestones, sandstones, and soils in the eastern United States	Can cause various cardiac, gastrointestinal, and neuromuscular diseases. Associated with hypertension and cardiotoxicity in animals

(continued)

Table 3.2 (continued)

Contaminant	Sources to groundwater	Potential health and other effects
Beryllium	Occurs naturally in soils, groundwater, and surface water. Often used in electrical industry equipment and components, nuclear power, and space industry. Enters the environment from mining operations, processing plants, and improper waste disposal. Found in low concentrations in rocks, coal, and petroleum and enters the ground	Causes acute and chronic toxicity; can cause damage to lungs and bones. Possible carcinogen
Cadmium	Found in low concentrations in rocks, coal, and petroleum and enters the groundwater and surface water when dissolved by acidic waters. May enter the environment from industrial discharge, mining waste, metal plating, water pipes, batteries, paints and pigments, plastic stabilizers, and landfill leachateMay be associated with the presence of sodium in drinking water when present in high concentrations. Often from saltwater intrusion, mineral dissolution, industrial and domestic waste	Replaces zinc biochemically in the body and causes high blood pressure, liver and kidney damage, and anemia. Destroys testicular tissue and red blood cells. Toxic to aquatic biota

(continued)

Table 3.2 (continued)

Contaminant	Sources to groundwater	Potential health and other effects
Chloride	May be associated with the presence of sodium in drinking water when present in high concentrations. Often from saltwater intrusion, mineral dissolution, industrial and domestic waste	Deteriorates plumbing, water heaters, and municipal water-works equipment at high levels. Above secondary maximum contaminant level, taste becomes noticeable
Chromium	Enters the environment from old mining operations runoff and leaching into groundwater, fossil fuel combustion, cement plant emissions, mineral leaching, and waste incineration. Used in metal plating and as a cooling-tower water additive	Chromium III is a nutritionally essential element. Chromium VI is much more toxic than Chromium III and causes liver and kidney damage, internal hemorrhaging, respiratory damage, dermatitis, and ulcers on the skin at high concentrations
Copper	Enters the environment from metal plating, industrial and domestic waste, mining, and mineral leaching	Can cause stomach and intestinal distress, liver and kidney damage, anemia in high doses. Imparts an adverse taste and significant staining to clothes and fixtures. Essential trace element but toxic to plants and algae at moderate levels
Cyanide	Often used in electroplating, steel processing, plastics, synthetic fabrics, and fertilizer production and from improper waste disposal	Poisoning is the result of damage to the spleen, brain, and liver

(continued)

Table 3.2 (continued)

Contaminant	Sources to groundwater	Potential health and other effects
Dissolved solids	Occur naturally but also enters the environment from man-made sources such as landfill leachate, feedlots, or sewage. A measure of the dissolved "salts" or minerals in the water. May also include some dissolved organic compounds	May have an influence on the acceptability of water in general. May be indicative of the presence of excess concentrations of specific substances not included in the Safe Water Drinking Act, which would make water objectionable. High concentrations of dissolved solids shorten the life of water heaters
Fluoride	Occurs naturally or as an additive to municipal water supplies; widely used in industry	Decreases incidence of tooth decay but high levels can stain or mottle teeth. Causes crippling bone disorder (calcification of the bones and joints) at very high levels
Hardness	Result of metallic ions dissolved in the water; reported as concentration of calcium carbonate. Calcium carbonate is derived from dissolved limestone or discharges from operating or abandoned mines	Decreases lather formation of soap and increases scale formation in water heaters and low-pressure boilers at high levels
Iron	Occurs naturally as a mineral from sediment and rocks or from mining, industrial waste, and corroding metal	Imparts a bitter astringent taste to water and a brownish color to laundered clothing and plumbing fixtures

(continued)

Table 3.2 (continued)

Contaminant	Sources to groundwater	Potential health and other effects
Lead	Enters the environment from industry, mining, plumbing, gasoline, coal, and as a water additive	Affects red blood cell chemistry; delays normal physical and mental development in babies and *young children. Causes slight deficits in attention* span, hearing, and learning in children. Can cause slight increase in blood pressure in some adults. Probable carcinogen
Manganese	Occurs naturally as a mineral from sediment and rocks or from mining and industrial waste	Causes aesthetic and economic damage and imparts brownish stains to laundry. Affects the taste of water and causes dark brown or black stains on plumbing fixtures. Relatively nontoxic to animals but toxic to plants at high levels
Mercury	Occurs as an inorganic salt and as organic mercury compounds. Enters the environment from industrial waste, mining, pesticides, coal, electrical equipment (batteries, lamps, switches), smelting, and fossil-fuel combustion	Causes acute and chronic toxicity. Targets the kidneys and can cause nervous system disorders

(continued)

Table 3.2 (continued)

Contaminant	Sources to groundwater	Potential health and other effects
Nickel	Occurs naturally in soils, groundwater, and surface water. Often used in electroplating, stainless steel and alloy products, mining, and refining	Damages the heart and liver of laboratory animals exposed to large amounts over their lifetime
Nitrate (as nitrogen)	Occurs naturally in mineral deposits, soils, seawater, freshwater systems, the atmosphere, and biota. More stable form of combined nitrogen in oxygenated water. Found in the highest levels in groundwater under extensively developed areas. Enters the environment from fertilizer, feedlots, and sewage	Toxicity results from the body's natural breakdown of nitrate to nitrite. Causes "blue baby disease," or methemoglobinemia, which threatens oxygen-carrying capacity of the blood
Nitrite (combined nitrate/nitrite)	Enters the environment from fertilizer, sewage, and human or farm-animal waste	Toxicity results from the body's natural breakdown of nitrate to nitrite. Causes "blue baby disease," or methemoglobinemia, which threatens oxygen-carrying capacity of the blood
Selenium	Enters the environment from naturally occurring geologic sources, sulfur, and coal	Causes acute and chronic toxic effects in animals—"blind staggers" in cattle. Nutritionally essential elements at low doses but toxic at high doses

(continued)

Table 3.2 (continued)

Contaminant	Sources to groundwater	Potential health and other effects
Silver	Enters the environment from ore mining and processing, product fabrication, and disposal. Often used in photography, electric and electronic equipment, sterling and electroplating, alloy, and solder. Because of the great economic value of silver, recovery practices are typically used to minimize loss	Can cause argyria, a blue-gray coloration of the skin, mucous membranes, eyes, and organs in humans and animals with chronic exposure
Sodium	Derived geologically from leaching of surface and underground deposits of salt and decomposition of various minerals. Human activities contribute through de-icing and washing products	Can be a health risk factor for those individuals on a low-sodium diet
Sulfate	Elevated concentrations may result from saltwater intrusion, mineral dissolution, and domestic or industrial waste	Forms hard scales on boilers and heat exchangers; can change the taste of water and has a laxative effect in high doses
Thallium	Enters the environment from soils; used in electronics, pharmaceuticals manufacturing, glass, and alloys	Damages kidneys, liver, brain, and intestines in laboratory animals when given in high doses over their lifetime
Zinc	Found naturally in water, most frequently in areas where it is mined. Enters environment from industrial waste, metal plating, and plumbing, and is a major component of sludge	Aids in the healing of wounds. Causes no ill health effects except in very high doses. Imparts an undesirable taste to water. Toxic to plants at high levels

Source:: (USGS, 2018)

Table 3.3 Microbiological contaminants found in groundwater

Contaminant	Sources to groundwater	Potential health and other effects
Volatile organic compounds	Enter the environment when used to make plastics, dyes, rubbers, polishes, solvents, crude oil, insecticides, inks, varnishes, paints, disinfectants, gasoline products, pharmaceuticals, preservatives, spot removers, paint removers, degreasers, and many more	Can cause cancer and liver damage, anemia, gastrointestinal disorder, skin irritation, blurred vision, exhaustion, weight loss, damage to the nervous system, and respiratory tract irritation
Pesticides	Enter the environment as herbicides, insecticides, fungicides, rodenticides, and algicides	Cause poisoning, headaches, dizziness, gastrointestinal disturbance, numbness, weakness, and cancer. Destroys the nervous system, thyroid, reproductive system, liver, and kidneys
Plasticizers, chlorinated solvents, benzo[a]pyrene, and dioxin	Used as sealants, linings, solvents, pesticides, plasticizers, components of gasoline, disinfectant, and wood preservative. Enters the environment from improper waste disposal, leaching runoff, leaking storage tank, and industrial runoff	Cause cancer. Damages nervous and reproductive systems, kidney, stomach, and liver

Source:: (USGS, 2018)

Table 3.4 Organic Contaminants found in Groundwater

Contaminant	Sources to groundwater	Potential health and other effects
Coliform bacteria	Occur naturally in the environment from soils and plants and in the intestines of humans and other warm-blooded animals. Used as an indicator for the presence of pathogenic bacteria, viruses, and parasites from domestic sewage, animal waste, or plant or soil material	Bacteria, viruses, and parasites can cause polio, cholera, typhoid fever, dysentery, and infectious hepatitis

Source:: (USGS, 2018)

Table 3.5 Physical Characteristics of Groundwater

Contaminant	Sources to groundwater	Potential health and other effects
Turbidity	Caused by the presence of suspended matter such as clay, silt, and fine particles of organic and inorganic matter, plankton, and other microscopic organisms. A measure of how much light can filter through the water sample	Objectionable for aesthetic reasons. Indicative of clay or other inert suspended particles in drinking water. May not adversely affect health but may need additional treatment. Following rainfall, variations in groundwater turbidity may be an indicator of surface contamination
Color	Can be caused by decaying leaves, plants, organic matter, copper, iron, and manganese, which may be objectionable. Indicative of large amounts of organic chemicals, inadequate treatment, and high disinfection demand. Potential for production of excess amounts of disinfection byproducts	Suggests that treatment is needed. No health concerns. Aesthetically unpleasant
pH	Indicates, by numerical expression, the degree to which water is alkaline or acidic. Represented on a scale of 0–14, where 0 is the most acidic, 14 is the most alkaline, and 7 is neutral	High pH causes a bitter taste; water pipes and water-using appliances become encrusted; depresses the effectiveness of the disinfection of chlorine, thereby causing the need for additional chlorine when pH is high. Low-pH water will corrode or dissolve metals and other substances

(continued)

Table 3.5 (continued)

Contaminant	Sources to groundwater	Potential health and other effects
Odor	Certain odors may be indicative of organic or nonorganic contaminants that originate from municipal or industrial waste discharges or from natural sources	
Taste	Some substances such as certain organic salts produce a taste without an odor and can be evaluated by a taste test. Many other sensations ascribed to the sense of taste actually are odors, although the sensation is not noticed until the material is ingested	

Source:: (USGS, 2018)

Table 3.6 Consolidated Water Quality Testing report of the water samples in DV Halli panchayat

Sl.no.	Village	Location of Source	Type of Source (BW-Surface well filter)	Name of Scheme (HP -MWS- PWS- MVS)	pH Water (013)			Turbidity (001)			Total Dissolved Solids (TDS) mg/L (015)			Nitrate (006)			Flouride (003)			Chloride (005)		
					6.5-8.5			NTU			500-2000			as NO3 ((mg-L)			as F (mg-L)			as Cl (mg-L)		
								1-5						45- N.R.			1.0- 1.5			250-1000		
					2019	2020	2021	2019	2020	2021	2019	2020	2021	2019	2020	2021	2019	2020	2021	2019	2020	2021
1	D. V Halli	Near Avikatte road	BW	RO			7.3			1			64			8			0.75			35
2	D.V Halli	Near Road side	BW	MWS			7.1			1			262			14			0.92			180
3	D. V Halli	Near Padagappa House	BW	MWS			7.5			1			410			16			1.42			135
4	D. V Halli	Near land	BW	OHT			7.5			1			393			14			1.1			155
5	Achenahalli	in the village	BW	OHT	7.27	6.8	7.9	1	0.05	1	675	435	272	22	10	10	1.02	1.75	1.3	110	210	142
6	Achenahalli	Near Guddaranganathaswamy	BW	MWS	7.28	6.2	7.3	0.05	1	1	729	748	188	31	19	13	1.75	1.75	1.1	180	292	108
7	Gurummana katte	Near kere	BW	MWS	7.24		7.5	1		1	752		196	22		12	1.23		1.2	215		102
8	Gurummana katte	Near Narasimiah House	BW	MWS	7.07		7.9	1		1	707		266	11		8	1.24		1.21	190		110
9	Thayegondanahalli	Near Hosaband	BW	MWS			7.4			1			298			20			1.3			104
10	Thayegondanahalli	Near Aralekatte	BW	MWS			7.2			1			340			18			1.2			123
11	R. N. Roppa	Near Maramma temple	BW	MWS	7.32		7.5	0.05		1	428		481	26		22	0.65		1.57	90		210
12	K.C. Roppa	Near Maramma temple	BW	OHT	6.8	6.9	7.3	1	1	1	860	878	633	38	14	25	1.23	1.95	1.59	225	175	288
13	Belladamadugu	Near Jayaramma house	BW	MWS	7.24		7.8	1		1	828		230	25		18	1.52		1.22	230		130
14	Somalara	In the village	BW	OHT			7.2			1			210			11			1.3			90
15	Somalara	Near Kote Rangappa temple	BW	MWS			7.4			1			163			10			1.11			98
16	Gundlahalli	Near Gowda house	BW	MWS	7.26		7.6	1		1	342		252	26		12	1.35		1.2	70		169
17	Gundlahalli	Near kere	BW	MWS	7.16		7.5	0.05		1	527		180	24		10	1.75		0.87	140		109

(continued)

Table 3.6 (continued)

Sl.no.	Village	Location of Source	Type of Source (BW- Surface well filter)	Name of Scheme (HP -MWS- PWS- MVS)	pH Water (013)			Turbidity (001)			Total Dissolved Solids (TDS) mg/L (015)			Nitrate (006)			Flouride (003)			Chloride (005)		
					6.5-8.5			NTU			500-2000			as NO3 ((mg-L)			as F (mg-L)			as Cl (mg-L)		
								1-5						45- N.R.			1.0- 1.5			250-1000		
					2019	2020	2021	2019	2020	2021	2019	2020	2021	2019	2020	2021	2019	2020	2021	2019	2020	2021
18	Gundlahalli	Near Kallapa land	BW	OHT	7.17		7.2	1		1	680		295	26		19	1.11		1.4	160		150

Sl.no.	Village	Location of Source	Type of Source (BW- Surface well filter)	Name of Scheme (HP -MWS- PWS- MVS)	Total Alkalinity (018)			Total Hardness (016)			Sulfate (007)			Total Iron (004)			Calcium		
					as CaCo2 (mg-L)			as CaCo3 (mg-L)			as SO4 (mg-L)			as Fe (mg-L)			as Ca (mg-L)		
					200-600			200-600			200-400			1.0-N.R			7.5-200		
					2019	2020	2021	2019	2020	2021	2019	2020	2021	2019	2020	2021	2019	2020	2021
1	D. V Halli	Near Avikatte road	BW	RO			22			30			12			0.03			12
2	D.V Halli	Near Road side	BW	MWS			210			200			110			0.07			80
3	D. V Halli	Near Padagappa House	BW	MWS			220			210			90			0.09			84
4	D. V Halli	Near land	BW	OHT			210			204			103			0.06			82
5	Achenahalli	in the village	BW	OHT	630	300	120	455	500	110	65	75	98	0.16	0.13	0.08	130	200	44
6	Achenahalli	Near Guddaranganathaswamy	BW	MWS	490	472	90	440	374	80	81	80	60	0.11	0.16	0.04	160	150	32
7	Gurummana katte	Near kere	BW	MWS	440		112	380		106	85		76	0.12		0.07	195		42
8	Gurummana katte	Near Narasimiah House	BW	MWS	520		140	450		128	78		77	0.12		0.06	130		50
9	Thayegondanahalli	Near Hosaband	BW	MWS			150			140			80			0.09			56
10	Thayegondanahalli	Near Aralekatte	BW	MWS			180			160			99			0.08			64
11	R. N. Roppa	Near Maramma temple	BW	MWS	430		248	310		240	48		124	0.14		0.06	65		96
12	K.C. Roppa	Near Maramma temple	BW	OHT	520	274	388	460	288	360	95	96	182	0.14	0.07	0.12	180	115	144
13	Belladamadugu	Near Jayaramma house	BW	MWS	550		128	530		118	92		100	0.14		0.08	90		46
14	Somalara	In the village	BW	OHT			108			102			86			0.09			40
15	Somalara	Near Kote Rangappa temple	BW	MWS			100			95			75			0.1			38
16	Gundlahalli	Near Gowda house	BW	MWS	360		130	250		119	38		94	0.14		0.06	50		48
17	Gundlahalli	Near kere	BW	MWS	488		132	330		128	58		82	0.16		0.07	135		48
18	Gundlahalli	Near Kallapa land	BW	OHT	520		149	480		139	75		105	0.16		0.06	135		56

References

Central Ground Water Board (2021, June) Dynamic groundwater resources of India 2020

Central Ground Water Board, Ministry of Jal Shakti, Department of Water Resources, RD &GR Government of India Report 2020

Davithuraj J, Manjunatha S (2014) Assessment of groundwater quality for drinking and irrigation purposes of upper Bennihalla Basin. Karnataka. Nat Environ Pollut Technol 13(1):57–62

Economic Times (2016, March 22) 76 million people in India have no access to safe water: Report—The Economic Times. https://economictimes.indiatimes.com/news/politics-and-nation/76-million-people-in-india-have-no-access-to-safe-water-report/articleshow/51513761.cms

Govardhann V (1990) Groundwater pollution hazards to human life-A case study of Nalgonda district.I.J.E.P. Vol.10 (1)54–61

Groundwater directorate, government of Karnataka, & central ground water board (2021, October) Dynamic Ground water resources of karnataka report 2020

Jaganmohan M (2021, May 6). India: population affected by groundwater contamination by contaminant 2020. Statista. https://www.statista.com/statistics/ Accessed 5 October 2021

Kulkarni C (2019, December 13) Groundwater getting toxic rapidly in Karnataka | Deccan Herald. https://www.deccanherald.com/state/karnataka-districts/groundwater-getting-toxic-rapidly-in-karnataka-785056.html

Kumar M, Puri A (2012) A review of permissible limits of drinking water.Indian J Occup Environ Med 16(1):40–44. https://doi.org/10.4103/0019-5278.99696

Misra AK (2014) Climate change and challenges of water and food security. Int J Sustain Built Environ 3(1):153–165. https://doi.org/10.1016/j.ijsbe.2014.04.006

Nanda N, Kumar P (2015, September 9) Ground water analysis in the surrounding areas of Madhugiri Taluk, Tumkur District, Karnataka, India., IJSR—International J Sci Res (IJSR), IJSR | World Wide JS, 4(9) 2277–8179

Raghavan S, Jayaram J, Lakshmisha A, Agarwal P, Nikam M, Nagendra B (2017) Advanced centre for integrated water resources management, Karnataka. 166

Ramakrishnaiah CR, Sadashivaiah C, Ranganna G (2009) Assessment of water quality index for the groundwater in Tumkur Taluk, Karnataka State. India. E-J Chem 6(2):523–530. https://doi.org/10.1155/2009/757424

Ramaswami V, Rajaguru P (1991) Ground water quality of Tiruppur. Ground Water Qual Tiruppur 33(2):187–191

Ravichandran S, Pundarikanthan NV (1991) Studies on ground water quality of Madras. Stud Ground Water Qual Madras 33(4):481–487

RDWSD Karnataka – Official Website. (Accessed 30 September 2021). https://english.swachhamevajayate.org/). Contamination of groundwater. USGS.gov | Science for a changing world.

USGS.gov (2018, October 9) Science for a changing world. https://www.usgs.gov/special-topic/water-science-school/science/contamination-groundwater?qt-science_center_objects=0#qt-science_center_objects

Vittala DSS (2017) Madhugiri taluk aquifer maps and management plan, Tumkur district, Karnataka. 17

Vyas HV, Sawant VA (2008) Seasonal variations in drinking water quality of some Borewell waters in urban area of Kolhapur city. Nat Environ Pollut Technol 7(2):261–266

Water Resources (2021, July 8) http://karenvis.nic.in/Database/Water_Resources_7898.aspx

Chapter 4
A Change Detection Analysis of Mangrove Forests in and Around Devi River Mouth, Odisha Using Remote Sensing and GIS Technique

Prasanna Kumar Nayak and Sujata Mishra

Abstract A mangrove is a type of forest species present in coastal parts of the World. Mangroves are present throughout the world in the tropic as well as sub-tropic region, mainly between latitudes 25° N and 25° S. The total mangrove forest area of the world surveyed in the early period of this decade is covered by 34,051,121.6 acres of land disseminated through more than above 100 countries and regions. Mangrove forests are one of the world's more susceptible ecosystems. Coastal Regulation Zone (CRZ) mapped the mangrove forests as Ecological Sensitive Zone. A lot of industrial development in the Coastal regions of the World as well as infrastructure growth leads toward the decreasing trend in the Mangrove ecosystem. The emphasis of this paper is to record the changes in mangrove ecosystem during 2006–2021 and the distribution of mangrove forests in and around Devi River Mouth of Odisha State, India. Devi River is one of the bifurcated river of Kathjodi and river Mahanadi which is also a part of Mahanadi Delta. It covers both Jagatsinghpur and Puri districts and finally falls into the Bay of Bengal. The potential of the study area for mangrove regeneration, having the Value of Longitudes and Latitudes 86°22′3.81″E to 86°24′56.18″E and 19°58′51.14″N to 20°2′35.33″N respectively.

Keywords Mangrove · Coastal · Zone · Regeneration · Tropical

The original version of the chapter has been revised: The missing references have been included. A correction to this chapter can be found at https://doi.org/10.1007/978-981-99-5479-7_19

P. K. Nayak (✉) · S. Mishra
S.C.S Autonomous College, Puri, Odisha, India
e-mail: prasnayak@gmail.com

P. K. Nayak
P.G. Department of Geography, Utkal University, Bhubaneswar, Odisha, India

A. Dubey et al. (eds.), *Promoting Sustainability Through Water Management and Climate Change Adaptation*, Advances in Geographical and Environmental Sciences, https://doi.org/10.1007/978-981-99-5479-7_4

4.1 Introduction

Mangrove plants are generally found in the very humid regions of the world. Where the plants occur is named as Mangrove ecosystem. Mangroves are more thriving and delicate. In the Odisha state, Mangrove forests are generally seen in the Mahanadi Delta region mainly in Devi River Mouth situated in both Jagatsinghpur and Puri districts and Bhitarkanika region in Kendrapara district.

Mangroves have provided a buffer zone to the local settlement between land and sea. Committees have adopted the environment and enabled a synchronized condition with Mangroves. People living close to mangrove areas with course of time have developed a feeling of respect to mangroves while using the resources available.

Mangrove ecosystem is highly fertile and insubstantial nature. Hence within a period of years management of mangroves is quite essential.

Remote sensing and GIS technique is a scientific approach to calculate the change in areas of mangrove plants that occur in the region. Mangrove maps prepared by remote sensing and GIS technique is very useful for the management of Mangrove ecosystem. Devi River is one of the bifurcated river of Kathjodi and river Mahanadi. It forges ahead *Jagatsinghpur* and *Puri district* in *Odisha* state and then falls into *Bay of Bengal*. Kathjodi River is the main distributary of river Mahanadi generated at *Cuttack*. Kathjodi River is renamed while forges toward east. Devi River passes through three blocks in Jagatsinghpur district and two blocks in *Puri district*. Finally it falls into Bay of Bengal after crossing 70 km from the south Mouth of River Mahanadi from the frontier line of Cuttack and Puri Districts. The mouth of the river is encompassed by concentrated forests having lack of population density.

4.2 Objective

The objective of this study is

1. To record the changes in mangrove ecosystems during last few years (2006–2021).
2. To study the recent situation and distribution of Mangroves in and around Devi River Mouth, Odisha.

4.3 Study Area

The study was conducted on mangroves in Devi River, which lie along the east coast of India between 19°58′51.14″N to 20°2′35.33″N latitude and 86°22′3.81″E to 86°24′56.18″E Longitude. Since Odisha belongs to coastal area, a large number of rare species like Bani (Avicenna), Rai (Rhizophora), Guan (Exocaria), and Sundari (Heritera), are available here. The sample area of mangroves including the mud flats is estimated between 2006 and 2021 over a period of fifteen (15) years to see whether the area will be increased or decreased.

4.4 20 Location Map

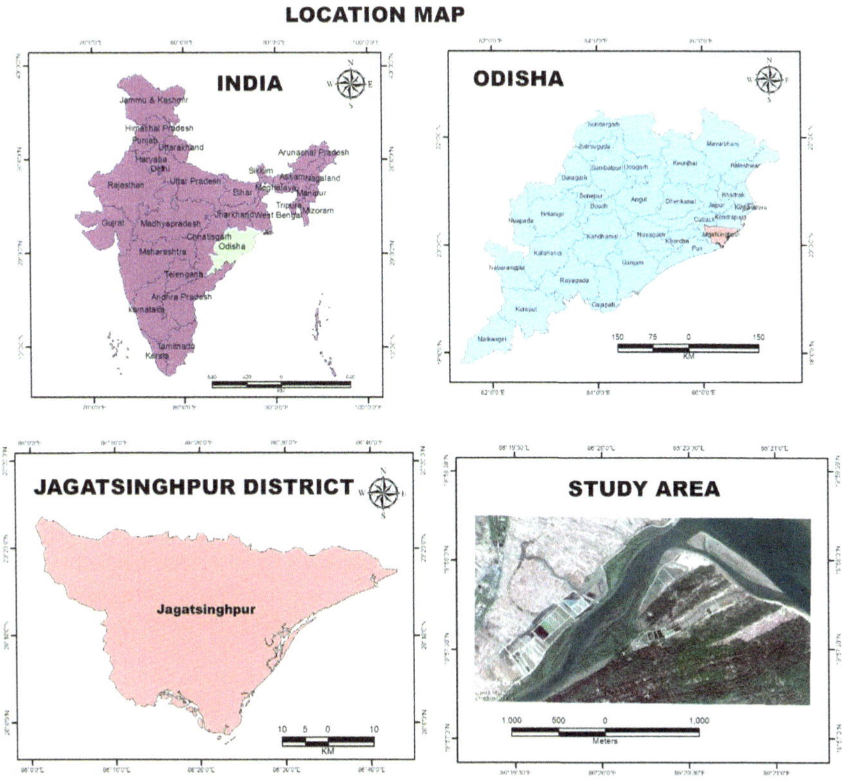

4.5 Methodology

In the present study, mangroves in Devi River mouth, Odisha were identified in High Resolution Satellite Imagery. Satellite image interpretation was carried out for the Change analysis of the mangroves identified within the study area. The Change detection analysis has been accomplished by the high resolution Google Satellite Imagery for the years 2006 to 2021 with a gap of 5 year period. The Shape files like Point, Line, and Polygons are digitized by using ArcGIS 10.0 and Google Earth Pro software. The mangroves shown in the study area generated through GIS Map in the year 2006 are cited as base year for the change detection analysis. The data

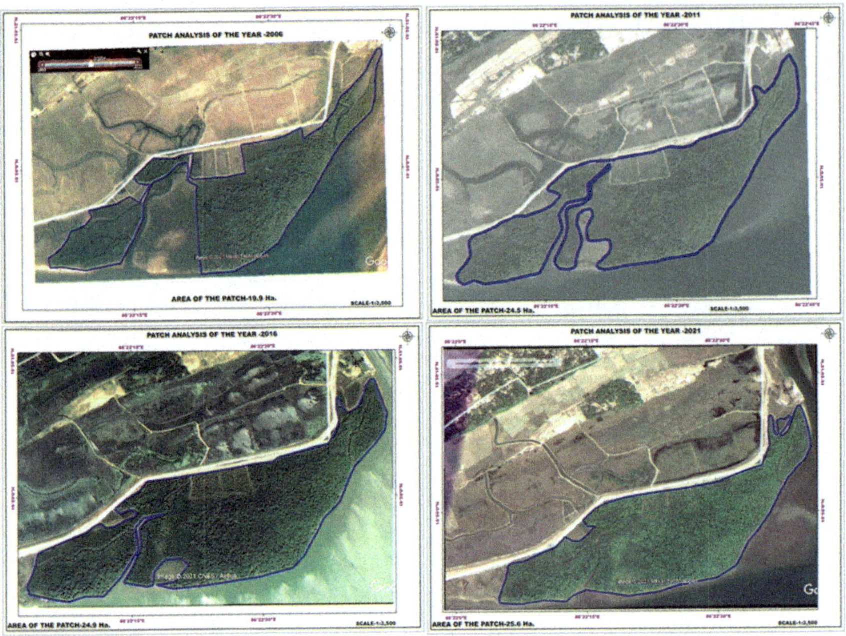

Fig. 4.1 Mangrove Areas depicted in Google Satellite Imagery in different years

quality generated in ArcGIS is validated through topology validation and ground truth mechanism. Similarly the change detection analysis has been carried out for the years 2011, 2016, and 2021. In the case of erroneous results, the change detection analysis of subsequent years is to be carried out and validated through above cited mechanism (Fig. 4.1).

4.6 Statistical Analysis

Sample area of mangroves taken in and around Devi River's mouth is depicted in the above-mentioned maps respectively. From 2006 to 2021, the area of mangroves increased, along the Devi River's mouth is depicted in Fig. 4.2 to 4 respectively. From the year 2006–2021, the area of mangroves increased in the study area from 19.9 to 25.6 Ha. The increasing trend is also depicted through Line and Pie Chart respectively.

In the study area, the increase in Mangrove forests leads toward a friendly atmosphere of fishery as well as reduces the risk of flood and storm. Apart from this in the Mouth area, the increase in Mangrove ecosystem checks the shoreline erosion profile and maintains the water quality. Further a healthy Mangrove forest protects the intervention of salt water into fresh ecosystem.

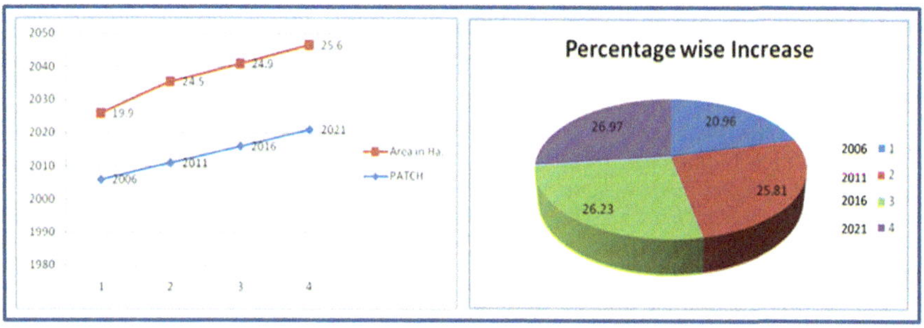

Fig. 4.2 Area of mangroves increased, along the Devi River's mouth

4.7 Conclusion

In accordance with the CRZ announcement, mangroves must be mapped in order to develop a Coastal Zone Management Plan. However, mapping alone will not aid in the appropriate implementation of CZMP. According to the change detection study, the area of mangroves has risen, but this increase should not be seen as a positive sign, as the increase has occurred on the seaward side and not the landward side, and there is no room for development of mangroves on the landward side. This was demonstrated by the charting of permanently lost and altered landscapes. This aided in the comprehension of the ecosystem's potential for Mangrove regeneration. Consequently, all of these characteristics must be taken into account during the local and national formulation of CZMP. Depending on whether plans should be developed for a particular location, local-level mapping can provide additional information.

4.8 Suggestions

a. In all deltaic regions, newly formed Mangrove forest blocks in UN-surveyed areas and in all islands/wetlands must be designated as Reserve forests. This proclamation would give legal protection against encroachment and irresponsible usage of these forest/sanctuary lands.
b. The destruction of all unauthorized prawn pens in mangrove/CRZ-I zones.
c. Extensive reforestation of damaged mangrove regions and riverbanks.
d. The provision of sufficient personnel, transportation, and communication facilities.
e. Preparation of a mangrove atlas based on digitized GIS maps of all mangrove regions that depict their actual land use pattern.
f. Elimination and prevention of mangrove invasion.
g. Preventing re-encroachment of recovered mangrove areas after eviction of trespassers.

References

Burrough PA, McDonnell RA (2000) Principles of geographical information, systems. Oxford University Press, New York

Gilman E, Ellison J, Duke NC, Field C (2008) Threats to mangroves from climate change and adaptation options: a review. Aquat Bot 89(2):237–250

Jayatissa LP, Dahdouh-Guebas F, Koedam N (2002) A review of the floral composition and distribution of mangroves in Sri Lanka. Bot J Linn Soc 138:29–43

Kairo JG, Dahdouh-Guebas F, Bosire J, Koedam N (2001) Restoration and management of mangrove systems—a lesson for and from the East African region. S Afr J Bot 67:383–389

Kaplan M, Reynaud FG, Lutcher's G (2009) Vulnerability assessment and protective effects of coastal vegetation during the 2004 Tsunami in Sri Lanka. Nat Haz Earth Syst Sci 9:1479–1494

Kathiresan K, Bingham BL (2001) Biology of mangroves and mangrove ecosystems. Adv Mar Biol 40:81–251

Komiyama A, Ong JE, Poungparn S (2008) Allometry, biomass and productivity of mangrove forests: a review. Aquat Bot 89(2):128–137

Krauss KW, Lovelock CE, McKee, KL, Lopez-Hoffman L, Ewe SM, Sousa WP (2008) Environmental drivers in mangrove establishment and early development: a review. Aquat Bot 89(2):105–127

Kristiansen E, Bouillon S, Dittmer T, Marchand C (2008) Organic carbon dynamics in mangrove ecosystems: a review. Aquat Bot 89(2):201–219

Lillesand TM, Kiefer RW, Chipman JW (2008) Remote sensing and image interpretation. John Wiley & Sons, New Delhi

Polidoro BA, Carpenter KE, Collins L, Duke NC, Ellison AM, Ellison JC, Farnsworth EJ, Fernando ES, Kathiresan K, Koedam NE, Livingstone SR, Miyagi T, Moore GE, Ngoc Nam V, Ong JE, Primavera JH, Salmo SG III, Sanciangco JC, Sukardjo S, Wang Y, Hong Yong JW (2010) The loss of species: mangrove extinction risk and geographic areas of global concern. PLoS One 5(4):e10095

Sabin's FF (1996) Remote sensing: principles and interpretation. W. H. Freeman and Company, San Francisco

Spalding M, Blasco F, Field C (1997) World Mangrove Atlas. The International Society for Mangrove Ecosystems, Okinawa, Japan, p 178

Taylor M, Ravilious C, Green EP (2003) Mangroves of East Africa. United Nations

Tomlinson PB (1986) The Botany of Mangroves. Cambridge University Press, Cambridge, UK

Triest L (2008) Molecular ecology and biogeography of mangrove trees towards conceptual insights on gene flow and barriers: a review. Aquat Bot 89(2):138–154

Walters BB, Rönnbäck P, Kovacs J, Crona B, Hussain S, Badola R, Primavera J, Barbier EB, Dahdouh-Guebas F (2008) Ethno biology, socio-economic and adaptive management of mangroves: a review. Aquat Bot 89(2):220–236

Chapter 5
Study of Spatio-Temporal Variation in Rainfall at Suketi River Basin by Using Rainfall Anomaly Index (RAI)

Ajay Kumar, Navneet Kaur, and Shilpa Devi

Abstract There is a great variation in the pattern of rainfall in any region, therefore it is important to study this phenomenon by using the appropriate methodology. In the present study Rainfall Anomaly Index (RAI) has been used to study the Spatio-temporal variation in the rainfall pattern at the Suketi river basin of Himachal Pradesh. Monthly rainfall data between the years 1985 and 2020 for two rainfall stations, i.e., Mandi and Sundernagar has been used. The rainfall Anomaly Index (RAI) is a useful tool to find out the rainfall pattern in terms of humid and dry years in any region. In this paper rainfall distribution pattern in the two most dry and humid years has also been calculated. The years with extreme rainfall events have also been analyzed.

Keywords Rainfall Pattern · Rainfall anomaly Index (RAI) · Suketi River Basin

5.1 Introduction

Decreasing per capita water availability due to variations in rainfall is a burning issue in today's world as rainfall directly affects water availability extremely. Conditions like floods and droughts are directly linked to the occurrence of rainfall. According to a report by Tripathi (2018), approximately 40% population of India will not have access to safe drinking water by 2030. If the population growth, urbanization, and exploitation rate of water resources continuously grow at the same pace, the prediction of NITI Aayog will become true before 2030.

Rainfall is the main source of water for both surface and groundwater resources. The rainfall trend analysis is important to understand the availability of water for utilization and management (Thenmozhi and Kottiswaran 2016). Various studies on rainfall trends find that the monsoonal rainfall variation causes climate change (Kundu et al. 2015). Agricultural activities also get affected by rainfall patterns,

A. Kumar (✉) · N. Kaur · S. Devi
Department of Geography, Panjab University, Chandigarh, India
e-mail: ajayberdhan@gmail.com

A. Dubey et al. (eds.), *Promoting Sustainability Through Water Management and Climate Change Adaptation*, Advances in Geographical and Environmental Sciences,
https://doi.org/10.1007/978-981-99-5479-7_5

particularly in the regions where irrigation facilities are not available. In a state like Himachal Pradesh, precipitation in the form of snow attracts tourists which is an important contributor to its economy, hence declining trend of precipitation directly affects the economy of the state.

Due to irregular rainfall and its concentration in specific seasons, monitoring rainfall by using climatic indices provides a better understanding of its patterns. On the basis of these indices, it is easy to study the characteristics of dry and rainy years in any physical unit (Djane Fonseca da Silva 2009). Silva (2015) used Rainfall Anomaly Index for the Mamanguape river basin in Brazil and found three different regions with precipitation patterns. The present study is an attempt to study the variation of rainfall in the Suketi river basin by using the Rainfall Anomaly Index. It will also highlight the pattern of rainfall in terms of rainy and dry periods between the years 1985 and 2020.

5.2 Data Sources, Methods, and Study Area

5.2.1 Data Sources

The present study is based on rainfall data which has been collected from the Indian Meteorological Department (IMD) for the period between 1985 and 2003 and the data from 2004 to 2020 has been collected from the Meteorological Centre, Shimla, Himachal Pradesh. In the basin, two rain gauges are placed at Mandi and Sundernagar. Monthly precipitation data has been used for the calculation of the Rainfall Anomaly Index (RAI). Annual RAI between 1985 and 2020 has been calculated to analyze the frequency and intensity of humid and dry years in the basin. Simultaneously, the monthly RAI has been calculated to analyze the rainfall distribution in the two driest and most humid years.

5.2.2 Methods and Statistical Analysis

In the present study Rainfall Anomaly Index (RAI) has been calculated to analyze the rainfall variation in the basin. To calculate the Rainfall Anomaly Index following formula has been used which has been developed by Rooy in 1965 and adopted by Freitas in 2005:

$$\text{RAI} = 3\left[\frac{N - \overline{N}}{M - \overline{\overline{N}}}\right] \text{ for Positive Anomalies}$$

$$\text{RAI} = -3\left[\frac{N - \overline{N}}{\overline{\overline{X}} - \overline{\overline{N}}}\right] \text{ for Negative Anomalies}$$

Table 5.1 Classification of rainfall anomaly index intensity

Range of Rainfall Anomaly Index	Classification
Above 4	Extremely Humid
2 to 4	Very Humid
0 to 4	Humid
−2 to 0	Dry
−4 to −2	Very Dry
Below −4	Extremely Dry

Source Freitas (2005), adapted by Araújo et al. (2009)

Here

$N =$ Current Month/Year/ Season in order when RAI will be calculated (mm).
$\overline{N} =$ Monthly/ seasonal/ annual average rainfall of the historical series (mm).
$\overline{M} =$ Average of the ten highest monthly/seasonal/ annual precipitation of the historical series (mm).
$\overline{X} =$ Average of the ten lowest monthly/seasonal/ annual precipitation of the historical series (mm).

Positive anomalies have values above average and negative anomalies have values below average. The rainfall Anomaly Index has been classified into five categories, i.e., Extremely Humid, Very Humid, Humid, Dry, Very Dry, and Extremely Dry (Table 5.1). In order to have a better analysis of rainfall anomalies in the basin the monthly RAI of the two driest and two rainiest years (extreme anomalies) has been analyzed.

5.2.3 *Study Area*

The present study has been conducted in the Suketi river basin of district Mandi in Himachal Pradesh (Map 5.1). It is located between 76°48'30" East to 77° East longitudes and 31°29' North to 31°45' North latitudes. It has an area of 422 square kilometers with an elevation of 754 m above mean sea level at Beas Suketi confluence and 2052 m above mean sea level at Zoomdhar, which is the highest peak of the basin. The average annual rainfall in the basin is approximately 1100 mm, out of which maximum rainfall is experienced in July and minimum in November. The basin is very significant in the entire Beas basin because of an intermontane valley which is called Bath valley. The Bath valley is known as the granary of Himachal Pradesh because of its fertility and agricultural production.

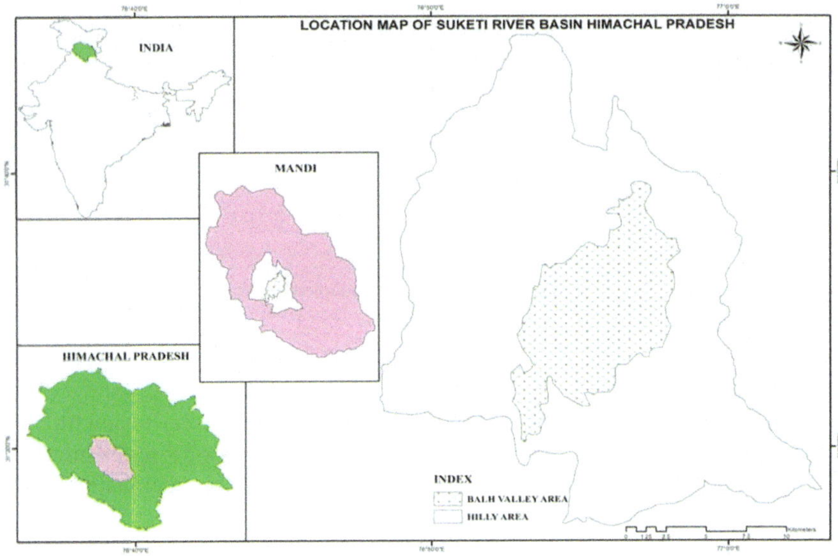

Map 5.1 Suketi river Basin: Location map of study area (*Source* Anil M Pophare and Umesh S Balpande 2014)

5.3 Results

The Rainfall Anomaly Index has been calculated for an annual and seasonal basis between 1985 and 2020. Annual rainfall of the basin between the same time period is 2933.62 mm., out of which 390.77 mm. (13.32%) happens in the pre-monsoon period, 1863.44 mm. (63.52%) in the monsoon period, 359.28 mm. (12.24%) in post-monsoon and 320.13 mm. (10.91%) in the winter season. The study shows that the annual rainfall in the Suketi river is anomalous. The average monthly rainfall during the study period has been shown in Fig. 5.1 which clearly depicts that maximum rainfall in the basin has been received between the months of June and September, which is known as the rainy or monsoon season. The rainfall in this period is much higher than the annual average rainfall of the basin. The highest rainfall occurs in the months of July and August and the minimum in October and November which is the driest period in the basin.

The months between December and February (winter season) shows dry spell with a small amount of rainfall in lower altitude and snowfall in the higher altitude of the basin. Between the months of March and May, which is the summer season very less amount of rainfall is experienced. Due to the small amount of rainfall, the increase in consumption of water and shrinking water bodies lead to the shortage of water in this season.

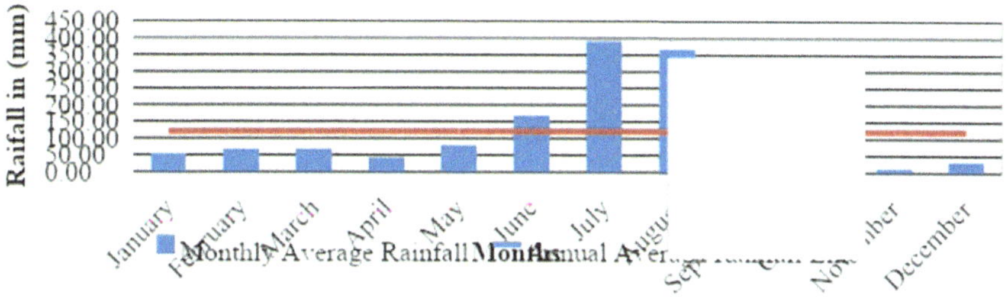

Fig. 5.1 Suketi River Basin: Average monthly rainfall, 1985–2020 (*Source* IMD and Meteorological Centre Shimla 2020)

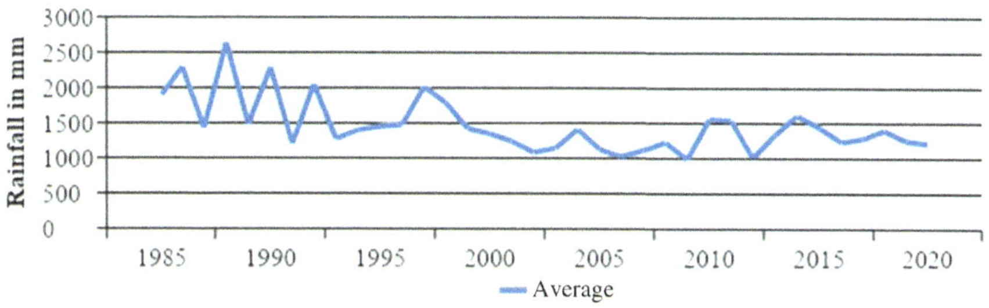

Fig. 5.2 Suketi River Basin: Annual average rainfall in Suketi river Basin, 1985 to 2020

5.3.1 Spatial Distribution of Rainfall in Suketi River Basin

A decreasing trend of rainfall has been observed in the basin between 1985 and 2020 (Fig. 5.2). The average annual rainfall of approximately 129.40 mm has been observed in Mandi station for the same time period. Pre-monsoonal average rainfall of this station is 68.77 mm, the monsoonal average rainfall is 335.81 mm, the post-monsoonal average rainfall is 60.40 mm, and winter rainfall average is 52.62 mm. It shows that the post-monsoon and winter season is the driest period at this station whereas the monsoonal season (June to August) is the most humid stretch in terms of rainfall.

On the other hand, the Sundernagar station shows an almost similar pattern of rainfall with few exceptions. The average annual rainfall of Sundernagar is approximately 115.7 mm for the same time period. The pre-monsoonal average rainfall of Sundernagar is 61.48 mm, the monsoonal average rainfall is 285.34 mm, the post-monsoonal average rainfall is 59.36 mm, and the winter average rainfall is 54.09 mm (Table 5.2). The detailed distribution of rainfall in these two stations has been discussed in the following sections:

Table 5.2 Suketi River Basin: Annual and seasonal average rainfall at Mandi and Sundernagar Stations

Seasons	Average Rainfall of Mandi Station between 1985 and 2020 (in mm)	Average Rainfall of Sundernagar Station between 1985 and 2020 (in mm)
Annual (January to December)	129.40 mm	115.7 mm
Pre-Monsoon/Summer (March to May)	68.77 mm	61.48 mm
Monsoon/Rainy (June to August)	335.81 mm	285.34 mm
Post-Monsoon/Autumn (September to November)	60.40 mm	59.36 mm
Winter (December to February)	52.62 mm	54.09 mm

Source: Compiled by Author

5.3.2 Distribution of Rainfall at Mandi Rain Gauge Station

Rainfall distribution is irregular in the Suketi river basin, where maximum rainfall has been observed in the monsoon season and minimum in the winter season. Maximum annual rainfall of approximately 3360 mm has been observed at Mandi station in the year 1988 and a minimum of 672.10 mm in the year 2006 and it continuously shows a declining trend till 2020 (Fig. 5.3). Maximum for pre-monsoon season, maximum (641.50 mm) rainfall has been recorded in the year 1987 and a minimum (37.30 mm) in the year 2006. Maximum rainfall (2404.60 mm) in the monsoon period has been recorded in the year 1986 and minimum (431.90 mm) in the year 2012. Post-monsoon rainfall was maximum (of 458 mm) in 1998 and minimum (of 11.40 mm) in the year 2020. Maximum rainfall (458.60 mm) for the winter season is recorded in the year 1990 and minimum (6.40 mm) in the year 2000. It is clear from the above discussion that there is a huge seasonal variation in the rainfall pattern at this station (Fig. 5.3).

5.3.3 Distribution of Rainfall at Sundernagar Rain Gauge Station

Rainfall distribution is irregular in the Suketi river basin, where maximum rainfall has been observed in the monsoon season and minimum in the winter season. Maximum annual rainfall of approximately 3360 mm has been observed at Mandi station in the year 1988 and a minimum of 672.10 mm in the year 2006 and it continuously shows a declining trend till 2020 (Fig. 5.4). Maximum for pre-monsoon season, maximum (641.50 mm) rainfall has been recorded in the year 1987 and a minimum (37.30 mm) in the year 2006. Maximum rainfall (2404.60 mm) in the monsoon period has been

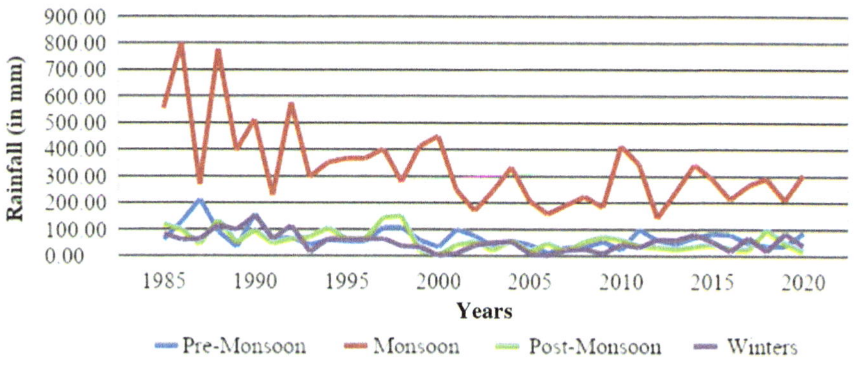

Fig. 5.3 Suketi River Basin: Seasonal rainfall at Mandi Station, 1985 to 2020

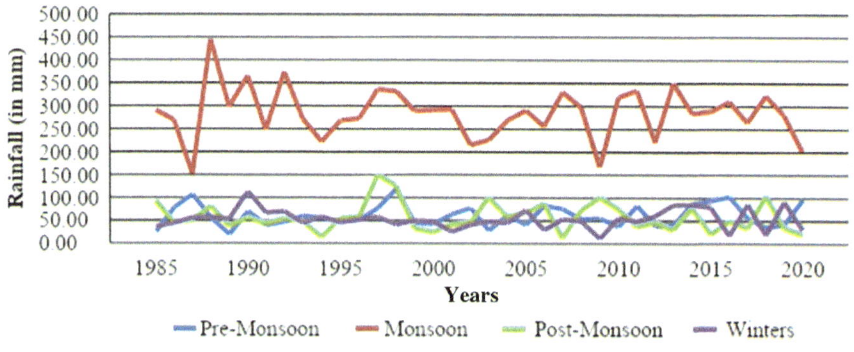

Fig. 5.4 Suketi river Basin: Seasonal rainfall at Sundernagar Station, 1985 to 2020 (*Source* Calculated on the basis of data from IMD Meteorological Centre Shimla 2020)

recorded in the year 1986 and minimum (431.90 mm) in the year 2012. Post-monsoon rainfall was maximum (of 458 mm) in 1998 and minimum (of 11.40 mm) in the year 2020. Maximum rainfall (458.60 mm) for the winter season is recorded in the year 1990 and minimum (6.40 mm) in the year 2000. It is clear from the above discussion that there is a huge seasonal variation in the rainfall pattern at this station (Fig. 5.5).

5.3.4 Rainfall Anomaly Index of Suketi River Basin

To analyze the rainfall distribution in the Suketi river basin, the Rainfall Anomaly Index has been calculated. The index highlights the dry and rainy years between 1986 and 2020. And simultaneously it also shows the intense and mild rainfall years. Positive RAI values have been recorded for 12 years which shows humid to extremely humid rainfall trends, however, 24 years have displayed negative RAI values which means they have dry to extremely dry rainfall. Results of the index indicate that the

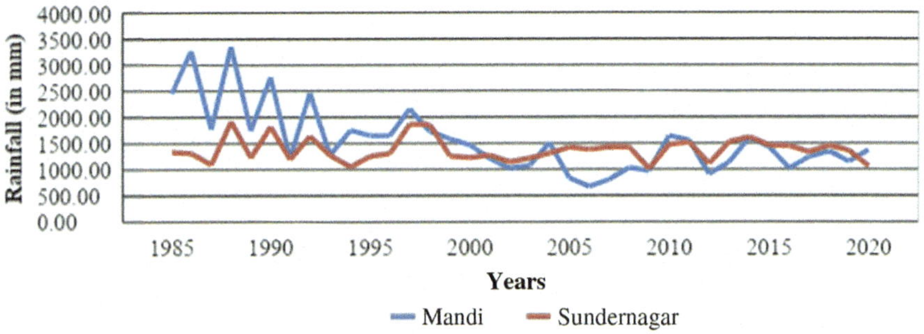

Fig. 5.5 Suketi river Basin: Rainfall variation at Mandi and Sundernagar Stations, 1985 and 2020

number of years with drought-like conditions is more than the number of humid years.

During the study period, the longest spell of dry years, i.e., 10 years have been recorded from 1999 to 2009. For the year 2009 highest negative RAI has been recorded (−4.06) which means this year had extreme dry conditions. On the other hand in the year 1988 highest positive RAI value has been recorded. This reflects extremely humid conditions during this particular year. During the last five years, i.e., 2015 to 2020 Suketi river basin is experiencing negative RAI value, which is the outcome of less rainfall and these trends can lead to drought-like conditions in near future.

The longest dry period of 10 years has been experienced between 1999 and 2009. The highest negative RAI value (−4.06) has been recorded in the year 2009 which indicates this year was an extremely dry year. On the contrary, 1988 was the year with a positive RAI value (6.98) indicating an extremely humid period. The duration of the last five years (2015 to 2020) shows a trend of negative RAI value which indicates that the basin is experiencing comparatively less rainfall which might be the potential cause of drying water sources in the basin.

To understand rainfall variability in a better manner Rainfall Anomaly Index for the two most humid and driest years has also been calculated. During the study period, the two most humid years were 1986 and 1988, however, the two driest years were 2009 and 2012. According to the results, the patterns of the two driest years were similar (Fig. 5.6), however, the pattern was also similar for humid years with an exception of May month's rainfall (Fig. 5.6).

5.4 Discussion

In the area of the great variation in precipitation, it is important to use an appropriate climate index to analyze its trend and scenario. In the present study, there is variation in the rainfall in terms of both space and time at the Suketi river basin. Therefore Rainfall Anomaly Index (RAI) has been used as a tool to study the rainfall variations

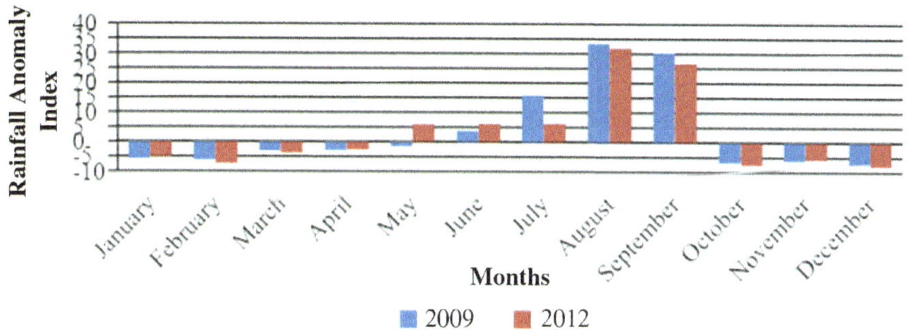

Fig. 5.6 Suketi river Basin: Monthly rainfall Anomaly index of two driest years (2009 and 2012) (*Source* Calculated on the basis of data from IMD Meteorological Centre Shimla 2020)

in the basin. The monthly rainfall data for two meteorological stations, i.e., Mandi and Sundernagar have been used for the years 1985 to 2020. From the collected data, months are divided into four seasons. These four seasons are the pre-monsoon or summer season (from March to May months), monsoon or rainy season (June to August), post-monsoon or autumn season (September to November), and winter season (December to February).

A total of 10 rainfall monitoring stations are located in district Mandi, out of which two stations namely Mandi and Sundernagar come under the Suketi river basin. A rain gauge at Mandi station has been installed in Mandi town near the confluence of the Beas and Suketi river. Another rain gauge has been installed at Sundernagar station which is near the origin of the Suketi river. The study shows that the rainfall in terms of frequency and intensity in these stations is uneven.

Although the actual distance between these two stations is about 30 km, there is a significant difference in rainfall distribution between them. In a comparison of rainfall in these two locations, it has been discovered that annual rainfall at Mandi station recorded approximately 2455.80 mm in 1985, whereas annual rainfall at Sundernagar station was approximately 1338.56 mm in 1985. For the year 1985, there is a difference of approximately 1117.24 mm in annual rainfall between these two locations. By the year 2020, the rainfall varies between these two locations have lessened to around 309 mm. In the year 2020 Mandi station received approximately 1360 mm of rainfall, while Sundernagar received approximately 1059.40 mm.

5.5 Conclusion

The rainy season in the Suketi river basin lasts from June to September, and the driest season is from October to December. From 1985 to 2020, twelve years have been recorded with a rainy trend and twenty-four years with a dry trend. During the study period, the most humid year was 1988, whereas the driest year was 2009. In comparison to the northern part of the basin, the southern part around Sundernagar receives less rainfall. The study also finds that the trend of rainfall is decreasing from the year of 1988 to 2020. The highest annual rainfall of 5283.12 mm has been observed in 1988. On the other hand rainfall of approximately 3225.60 mm has been observed in 2014 which is the highest in the last twenty years, i.e., 2000 to 2020. The declining trend of rainfall also affects the availability of water resources; therefore there is a need to increase the potential of groundwater recharge. The surface water availability can be increased by reducing the wastage of rainfall water flow by structural and non-structural methods, hence the sustainability of water resources can be ensured.

References

Araujo LE MN (2009) Classificação da precipitação anual e da quadrachuvosa da bacia do rio paraíba utilizando índice de anomalia de chuva. Rev.Ambient. Água., 93–110

Araújo LE, Moraes Neto JM, Sousa F (2009) Classificação da precipitação anual e da quadra chuvosa da bacia do rio paraíba utilizando índice de anomalia de chuva (IAC). Rev. Ambient. Água. 4(3):93–110

Balpande US (2014) Morphometric analysis of Suketi river basin, Himachal Himalaya, India. J Earth Syst Sci 123(7):1501–1515

Djane Fonseca da Silva Fd (2009) Influence of global climate variability and its temporal scales on the precipitation In Alto Mundaú (PE). Rev. Bras.Geogr. Fís., 64–82

Freitas M (2005) Um sistema de suporte à decisão para o monitoramento de secas meteorológicas em regiões semiáridas. Rev Tecnol (suppl 19):84–95

Kundu SK (2015) Analysis of spatial and temporal variation in rainfall trend of Madhya pradesh, India (1901–2011).Environ Earth Sci, 8197–8216

Mas F (2005) Um sistema de suporte à decisão para o monitoramento de secasmeteorológicas em regiões semi- áridas. Rev. Tecnol., 84–95

Sati DRO (1998) Handbook on watershed management. Society of people for development, 45

Silva EB (2015) Precipitation regionalization of the Brazilian Amazon. Rev. Bras. Eng. Agríc. Ambient, 9–14

Singh HP (2010) Impact of drought on water resources of Himachal Pradesh. Biol Forum Int J, 73–77

Thenmozhi MA (2016) Analysis of rainfall trend using mann-kendall test and the sen's slope estimation in Udumalpet of tirapur district in tamilnadu. Int J Agric Sci Res, 131–138

Tripathi B (2018, June 25) 40% of Indians will have no access to drinking water by 2030: NITI Aayog. Business Standard

Chapter 6
Hydrogeomorphic Investigation to Select the Suitable Locations for Water Conservation Structures in PG-4 Watershed of Painganga River Basin of the Buldhana District, Maharashtra, India

Sandip K. Sirsat, Mohan A. Sonar, and Vishranti B. Kadam

Abstract This study presents our remote sensing technique data on PG-4 watershed of the Painganga River of Buldhana district, Maharashtra, India, in order to delineate the hydrogeomorphic unit. Our analysis suggests that the groundwater potential of the hydrogeomorphological units including plateau moderately dissected, plateau slightly dissected, plateau weathered, and plateau weathered shallow are found to be in a good to moderate state, whereas two other units including Plateau un-dissected and Escarpment slope exhibit poor to moderate groundwater potential. The maximum area of the studied watershed exhibits low drainage (0 to 2.31 km), which is good for the percolation and recharge of the groundwater. The studied area also shows the lowest slope value (0–1.68), so that it can be considered for the construction of water conservation structures including those to support the groundwater resources. In the PG-4 watershed, ~75% area is nearly flat and hence it is good for the construction of the groundwater recharge structures. These water conservation structures are also suitable for the proper management of soil and surface water resources. We have recommended 78 locations for the artificial structures in the PG-4 watershed, out of them 38 locations are suitable for nala plugging, 28 for percolation tanks, and 12 for check dams.

Keywords Groundwater · GIS · Geomorphic unit · PG-4 watershed · Conservation structure

---

S. K. Sirsat (✉) · M. A. Sonar
Department of Geology, Government Institute of Science, Aurangabad, Maharashtra 431004, India
e-mail: sksirsat@gmail.com

V. B. Kadam
Department of Geology, Deogiri College, Aurangabad, Maharashtra 431005, India

A. Dubey et al. (eds.), *Promoting Sustainability Through Water Management and Climate Change Adaptation*, Advances in Geographical and Environmental Sciences,
https://doi.org/10.1007/978-981-99-5479-7_6

6.1 Introduction

Water and soil are among the most dynamic natural resources of the earth on which life is intimately dependent. The development and evolution of life also depend on these natural resources. These resources are limited and being increasingly exploited day by day due to the increasing burden of the rising human population. This prone a serious threat to the preservation and sustainable development of these natural resources. Therefore, the proper use of these limited resources is very important to ensure their safe carry forward to the next generation. The proper use of these resources can be done through water and soil resource planning, conservation, and management (Choudhari et al. 2018). For example, watershed management plays an important role in the conservation of water and soil resources and their sustainable development (Jadhav et al. 2009; Choudhari et al. 2018).

Poor conservation and management (both in terms of exploitation and pollution) of the surface water resources such as rivers, lakes, and ponds, day by day there is a continuously growing usage of the groundwater to fulfill the daily domestic, agricultural, and industrial demand (Siebert et al. 2010; Zaveri et al. 2016; Ramaiah et al. 2012; Rajaveni et al. 2017; Selvakumar et al. 2017; Shinde et al. 2020). Therefore, the groundwater potential of every watershed has to be monitored and properly conserved for a sustainable future. The groundwater potential of the alluvial or soft-rock area is likely to be in good condition, however, the potential of hard rock terrains is always been problematic and their understanding needs the knowledge of geology as well as geomorphology of the area. In terms of geology and geomorphology the groundwater potential of the area is directly or indirectly controlled by several factors such as terrain features such as weathering rate, the number of fractures, permeability, slope, drainage pattern, landforms, land use/land cover, and climate (Jaiswal et al. 2003, Surrette et al. 2008; Ravindran & Jeyaram 1997; Lokesh et al. 2007; Sidle & Onda 2004; Babar 2005; Kudrna & Sindelarova 2006). Generally, there is always a possibility of finding groundwater in hard rock terrain mostly confined to the fractured and weathered horizons, although to a limited extent (Kumar & Kumar 2010).

Relation between geomorphology and groundwater often approaches other developing technical fields, such as hydroecology or hydrogeology (Loague et al. 2006; Hancock et al. 2009). Geomorphic units are important in the identification and classification of several landforms for groundwater management and development activities in a particular watershed (Muthamilselvan 2017). GIS and remote sensing are capable of covering an inaccessible and vast area of the Earth's surface for unlimited periods and one can extract surface area information multiple times for any season that could help in determining the areas of groundwater potential and identifying sites for the artificial recharge (Senthilkumar et al. 2019). Although the groundwater potential of any area cannot be fully understood directly from the remotely sensed data, the use of remote sensing must be used to infer in the identification of surface features that act as an indicator of high or low groundwater potential (Das et al. 1997; Ravindran and Jeyaram 1997). Hydro geomorphological mapping of the groundwater prospect zones is important as it provides the current spatial disposition of basic information on for example geology, surface water bodies, landforms, and soil that can reveal the groundwater movement and reorganization (Murthy 2000).

Artificial groundwater recharge can be very helpful in improving the groundwater potential of the area and was generally been conducted in the following: (i) regions with groundwater levels shrinking regularly, (ii) regions with most of the aquifers are already been desaturated/destroyed, (iii) regions with the accessibility of groundwater is inadequate in the months of scarcity, and (iv) regions with excess salinity in the groundwater (Chandramohan et al. 2019). The main objective of the present study is to identify the groundwater potential area and select sites for the proposed water conservation structures in the PG-4 watershed of the Painganga river basin.

The given PG-4 watershed is occupied by an area of 196.76 sqkm in the Buldhana district of Maharashtra (Fig. 6.1). The coordinates for the area are 200 1′ 54″ N to 200 9′ 51″ N and 760 21′ 52″ E to 760 36′ 00″ E. Rocks in the studied basin are dark-grey colored, fine to medium-grained, massive porphyritic basalts of Deccan Traps of Upper Cretaceous to Lower Eocene age (Deshpande 2012; CGWB 2019). These basaltic lava flows occupy ~94% of the total area, which comprises a thick pile of lava flows, which are infrequently divided by thin seams of red-bole. The lava flows are of both 'aa' and 'pahoehoe' type (Gurav and Babar 2019; Madhnure and Tambe 2012; Madhnure 2014). These two types of lava flow exhibit different hydrogeological characteristics. The PG-4 watershed shows a well-developed dendritic to sub-dendritic drainage pattern, which is largely controlled by relief, homogeneous lithology, and nearly flat and rolling topography (Babar et al. 2011; Shimpi et al. 2014). The study area is largely covered by black cotton soil, which is substantial for agricultural purposes (ESR 2016).

6.2 Data and Methods

A hydrogeomorphic analysis is done by using remote sensing and GIS techniques. SRTM-DEM data with a 90 m resolution was downloaded from USGS (https://earthexplorer.usgs.gov) and the DEM map was created by using the software Arc GIS 10.3. The extractions of the stream network were prepared using the hydrology tool and a slope map was prepared using the DEM data in Arc GIS 10.3 software. Geomorphological map data was downloaded from Bhukosh, Geological Survey of India. Different water conservation structures including nala plugging, percolation tank, and check dam were studied using drainage and slope of the ground in ArcGIS10.3 software is also used for the generation and preparation of thematic data or maps.

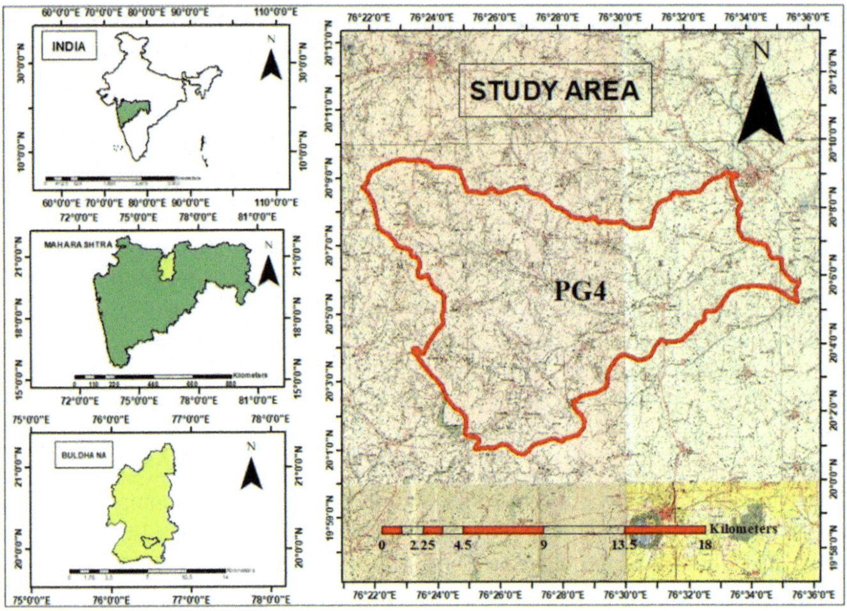

Fig. 6.1 Location map of PG-4 watershed

6.3 Results and Discussion

The geomorphological map of the Plainganga river basin is shown in Fig. 6.2 and the geomorphic unit is in Table 6.1. Various abbreviations used in the table include ES—Escarpment slope, PLM—Plateau moderately dissected, PLS—plateau slightly dissected, PLU—Plateau un-dissected, PLW—plateau weathered, and PLWS—plateau weathered shallow.

6.3.1 Hydrogeomorphology

The term 'hydrogeomorphology' designates the study of landforms in direct relation to the occurrence of groundwater (Todd 1980; Babar 2005). Hydrogeomorphic units play an important role in identifying the groundwater potential of different areas (Rani et al. 2015). A combined unit that includes structure, landform, lithology, and recharge conditions is called a hydrogeomorphic unit. Such units are measured as three-dimensional homogenous units concerning recharge conditions, hydrogeological characters, and they are also preserved as aquifers. The groundwater prospects are predicted to be uniform in a hydrogeomorphic unit. To study a hydrogeomorphic unit for groundwater prospects, an inventory of the monitoring factors such as rock type, landform, structure, and recharge condition, has to be understood and their

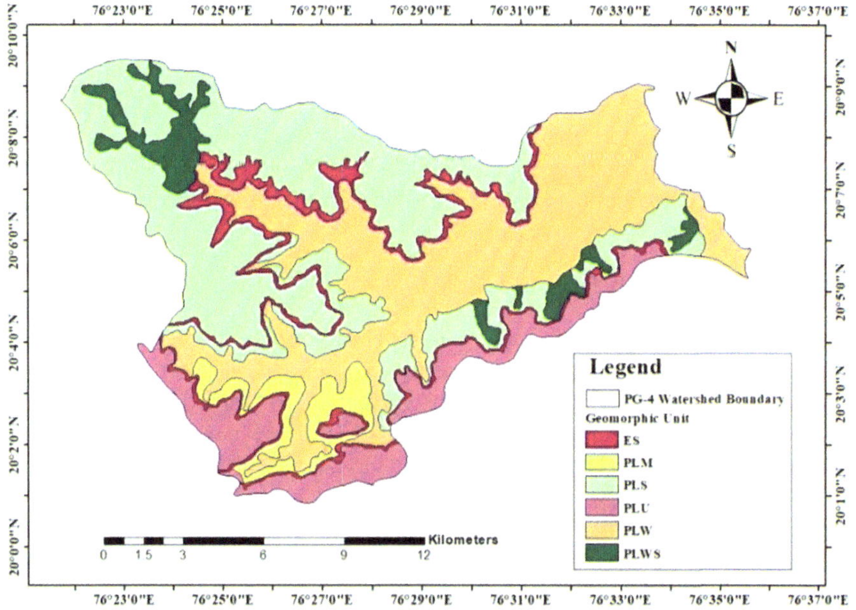

Fig. 6.2 Geomorphological map of PG-4 watershed

Table 6.1 Area covers by different geomorphic units of PG-4 watershed

Geomorphic unit	Area km^2	Water potential
Escarpment (Es)	14.57	Poor
Plateau moderate dissected (PLM)	8.51	Moderate
Plateau slightly dissected (PLS)	75.98	Good
Plateau un-dissected (PLU)	20.79	Very poor
Plateau weathered (PLW)	66.63	Good
Plateau weathered shallow (PLWS)	10.28	Moderate
Total	196.76	-

hydrogeological characteristics need to be evaluated. For example, in a hydrogeological view, groundwater in a basaltic terrain occurs generally in the upper weathered and fractured zones, and potential zones may occur at deeper levels in the form of fractures and inter-flow zones (Shimpi et al. 2014).

The surface lithology and structural features have the main control over the occurrence of natural water resources (Shimpi et al. 2014). Geomorphic units are demarcated based on image classifications such as tone, texture, shape, color, and association of features of remotely sensed data. In the present study we have found six geomorphic units from the satellite imagery (Fig. 6.2) and the groundwater potential of each unit is explained below.

i. Escarpments (Es)

Escarpments have steep slope that forms as a result of erosion and divides two different zones with different elevations. In an escarpment generally, one side is steeper and eroded more than the other, which is resulted from unequal surface erosion. Steep slopes that are augmenting the runoff zones have poor potential for groundwater occurrences (Singh et al. 2013). Escarpment is covered by 14.57 sqkm area in the studied PG-4 watershed.

ii. Plateau Moderately Dissected (PLM)

This type of geomorphic unit represents a plateau dissected by valleys or gullies sand is generally formed due to differential erosion and weathering. These areas have steep to moderate slopes (Gaurav and Babar 2019). These units cover around an area of 8.51 sqkm in the studied watershed, which shows a moderate potential for groundwater resources. This geomorphic unit has a moderate infiltration or recharge rate through the primary hydrological features and its storage capacity is largely influenced by the secondary features. The shallow aquifers in the area partially drain out into the deep valleys.

iii. Plateau Slightly Dissected (PLS)

Plateau slightly dissected is a kind of plateau that is dissected by drainages and its original form is considerably changed (Sharma and Shukla 2015). This unit covers around 75.98 sqkm area in the studied PG-4 watershed Plainganga river basin. This unit is inferred to have a good recharge and storage zone for groundwater. Plateau slightly dissected units are good potential zones for groundwater resources or groundwater explorations along the valley portion and recharge depends on its relative elevation compared to the nearby landforms.

iv. Plateau Un-dissected (PLU)

This unit leads in the total recharge of the watershed and it covers a 20.79 sqkm area in the southern part of the watershed and lies between the runoff and storage zone. This unit is inferred to be a very poor potential for groundwater resources.

v. Plateau Weathered (PLW)

This unit is lying parallel to the stream course and is dominant in the middle and northeast parts with a gentle slope of the watershed. Weathered plateaus are either recharge or flood plains zone. It covers around 66.63 sqkm area in the PG-4 watershed. This unit is inferred to be good potential for groundwater resources with good recharge and storage zones depending upon the thickness of weathering and accumulated material supportive for recharge.

vi. Plateau Weathered Shallow (PLWS)

This unit is present mostly in the northern part of the watershed and also covers some areas in the southeast. It is a recharge or storage zone that covers around 10.28 sqkm area. The groundwater is usually available in the monsoon and winter seasons only and remains dry during the summer season; therefore, this unit is considered to have a moderate potential for groundwater resources.

6.3.2 Identify Suitable Sites for Water Conservation Structures

Remote sensing and GIS techniques can be very helpful while attempting to prioritization of river basins and found to be suitable sites for groundwater resources or different water conservation structures. The drainage and slope map we have prepared to find out a suitable location for nala plugging, percolation tanks, and check dams.

Drainage patterns denote spatial relationships with the stream or rivers, which may be influenced in their erosion by variations in the slope, rock resistance, structure, and geologic history of a region. The dendritic drainage pattern is associated with areas having less structural control, homogeneous lithology, and very gentle or flat rolling topographic surface (Shimpi et al. 2014). The streams follow litho-contacts but places do not coincide with the lithological boundaries (Sharma and Shukla 2015). Deccan traps are fractured and jointed undulated topography with dendritic to sub-dendritic drainage patterns (Khanday and Javed 2017). The dendritic to sub-dendritic drainage pattern is observed in the watershed (Fig. 6.3). Different streams of drainage networks were extracted and delineated from Digital Elevation Model (DEM) with the assistance of a special analyst tool in the Arc tools box of ArcGIS 10.3 software. Extracted drainage networks were overlapped on the digitized stream of the Survey of India Top sheet for cross-validation. In the PG-4 watershed are observed maximum first-order and second-order streams which are associated with relatively higher elevations where relief and slope are high resulting in more runoff and less recharge (Khanday and Javed 2017). We also observed third, fourth, and fifth-order streams having lower elevation as compared to first-order and second-order streams resulting in less runoff and high recharge (Fig. 6.4).

Drainage density shows the distances among streams and provides the length of streams within one unit area (Sharma and Shukla 2015). High drainage density is suggesting of high surface runoff and low drainage density suggests low groundwater recharge. High drainage density indicates an increase in food peaks, whereas there is a decrease in food levels in low drainage density (Pallard et al. 2009). The drainage density map (Fig. 6.5) shows that it ranges between 0 and 11.58 km and the class 0 to 2.31 km range occupies a maximum area of the PG-4 watershed, hence it is good for groundwater recharge and percolation of water.

The slope is one of the main factors for monitoring the residence time of water runoff and thereby the percolation of groundwater into the subsurface of any terrain (Muthamilselvan 2017). The slope has a leading impact on the surface and subsurface

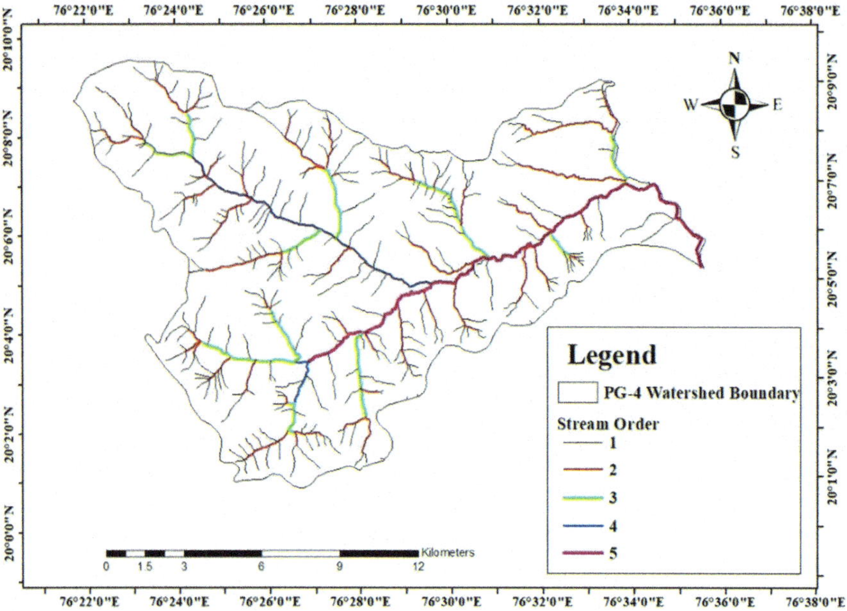

Fig. 6.3 Drainage map of PG-4 watershed

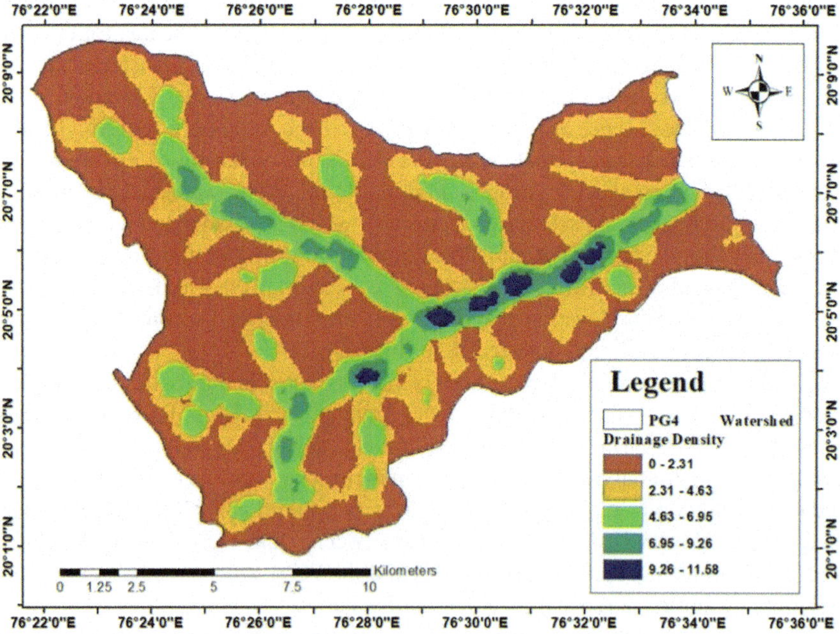

Fig. 6.4 Drainage density map of PG-4 watershed

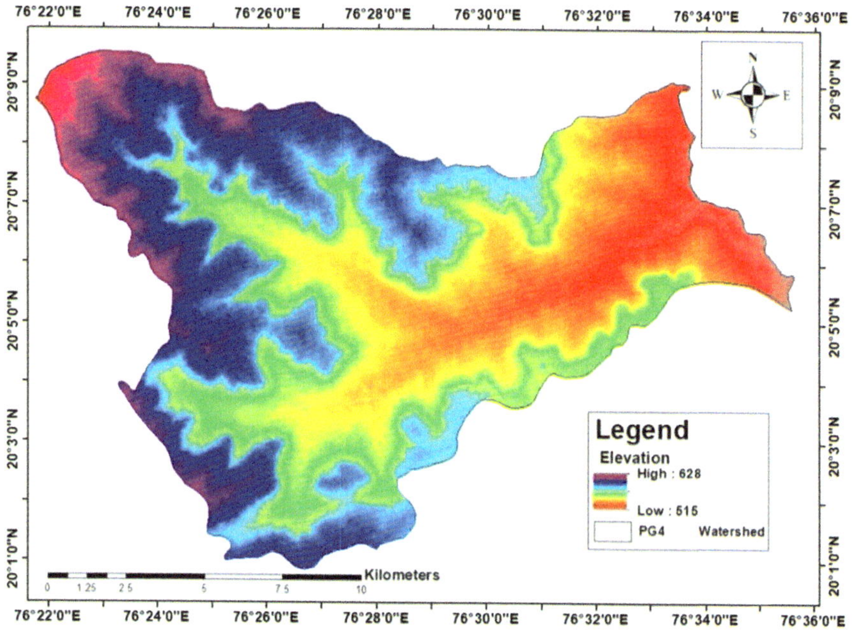

Fig. 6.5 Digital elevation model of PG-4 watershed

water flows and manages the water percolation or infiltration rate, erosion intensity, and surface runoff. So, it plays a dynamic role in providing information about suitability for groundwater prospects and sitting recharge structures. The slope is gentle in that area surface runoff is slow and also leads to more infiltration for rainwater to percolate in these regions. An appropriate slope for making artificial recharge structures in too flat slope would affect the drainage runoff and a too steep slope would make it difficult to construct and maintain the structure. Digital elevation model (DEM) of the study area showing elevation variation in the study area (Fig. 6.5). The maximum elevation of the study area is 682 m and the minimum elevation is 515 m.

The slope of the area is derived from ASTER DEM 90 m resolution and it varies from 0 to 8.4 degrees which is shown in Fig. 6.6. The slope is classified into six classes of the area based on the method given by Patil and Mohite (2014). The major part of the area is having a gentle or flat slope and this slope is less moderate for groundwater prospects. However, variations in the slope (0 to 8.4) were made to group the entire area into six slope classes, i.e., 0–1.68, 1.68–3.36, 3.36–5.04, 5.04–6.72, 6.72–8.4, and >8.4 percent and it occupied 145.70 sqkm, 34.10 sqkm, 13.78 sqkm, 2.86 sqkm, 0.28 sqkm, 0.04 sqkm area respectively (Fig. 6.6). Whereas, major part of the study area is with gentle or flat slope occupying 0 to 1.68 (74.05%) area of the watershed; hence it is good for groundwater recharge and percolation of water.

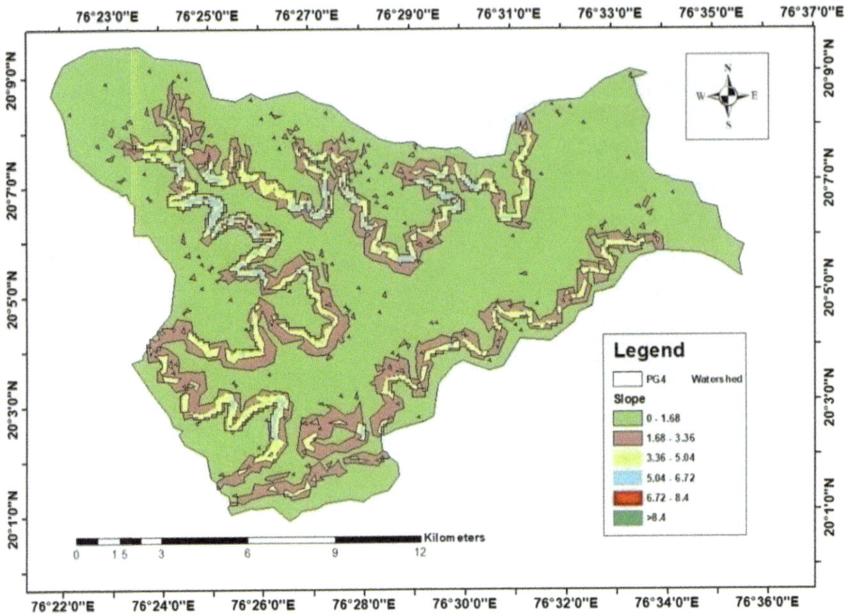

Fig. 6.6 Slope map of PG-4 watershed

6.3.3 Groundwater Resources or Artificial Recharge Structures

Based on topography, lithological characters, slope, drainage pattern or stream ordering and groundwater prospecting zones and GIS modeling has been supported for artificial recharge structures such as nala plugging, percolation tanks, and check dams (Fig. 6.6).

Totally, 78 conservation structure locations were suggested in PG-4 watershed, out of these 38 locations for nala plugging, 28 locations for percolation tank, and 12 locations for check dam (Fig. 6.6). It is suitable for the proper groundwater management practices and these studies are important for the implementation of soil erosion prevention practices as well as soil management observations (Choudhari et al. 2018).

Nala plugging structures are a good influence on the recharge of water and the high infiltration rate of aquifers. Nala plugging is constructed across the lower order streams mostly in first-order streams (Table 6.2) having a gentle slope and is feasible both in hard rock as well as alluvial formation to recharge and store the water.

Percolation tank is constructed across the middle-order streams such as second- or third-order streams (Table 6.2) for storing water increasing water percolation and improving soil moisture regimes. Percolation tanks are the most predominant structures in the deliberate watershed as an amount to recharge the groundwater reservoir. The efficacy and feasibility of these structures is more in hard rock terrain

Table 6.2 Preferred artificial structure using stream ordering in PG-4 watershed

Stream orders	Total streams	Preferred artificial structures
1st order	260	Nala Plugging
2nd order	48	Percolation Tank
3rd order	12	Percolation Tank
4th order	03	Check Dam
5th order	01	Check Dam
Total	324	–

where the rocks are highly weathered and fractured (Shimpi et al. 2014). The selection criteria for recharge through percolation tanks like the occurrence of weathered and fractured basalt and areas having gentle slope are eligible (Fig. 6.7).

In hard rock as well as in alluvial formation check dams are common structures constructed across small streams having gentle slope. The site selected for check dam should have appropriate width of the permeable bed or weathered formation to allow recharge of stored water within a short period. Check dams are constructed across higher order streams like fourth- or fifth-order streams (Table 6.2) in areas having a gentler slope. For demarcation of zones suitable for construction of check dam criteria like the presence of weathered and fractured basalt or alluvium; areas are flowing by higher ordered streams; areas below agricultural practices or wasteland surface; areas having good, moderate, moderate to poor groundwater prospects and

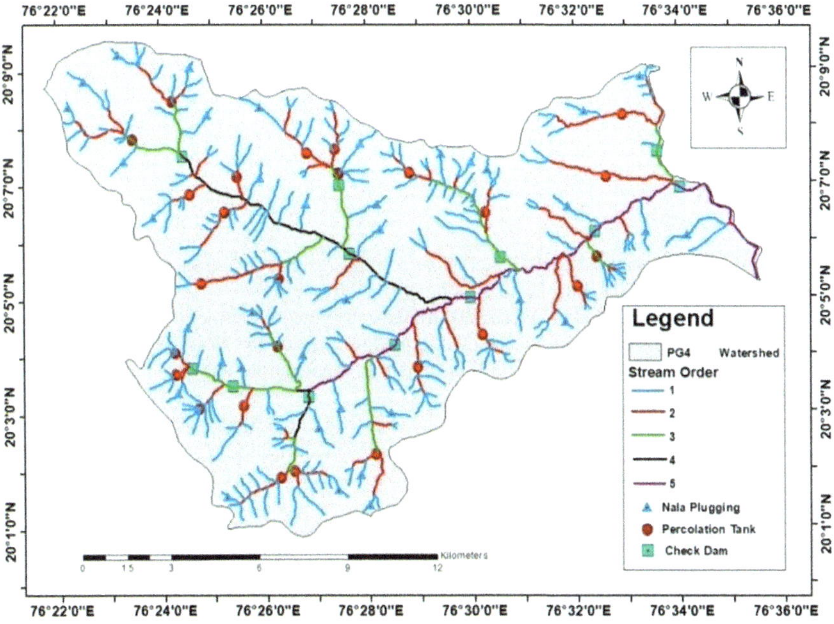

Fig. 6.7 Preferred artificial structure sites in PG-4 watershed

areas having gentle slope are eligible for recharge trough check dams. The check dam sites suggested for the PG-4 watershed are shown in Fig. 6.7.

6.4 Conclusion

A hydrogeomorphological investigation supported by the GIS mapping method is a very useful tool in the assessment of infiltration potential. The hydrogeomorphic units like they have good groundwater potential PLM, PLS, PLW, and PLWS have moderately to good groundwater potential. The hydrogeomorphic units like plateau un-dissected (PLU) and escarpment (ES) have poor groundwater potential zone. Dug wells are recommended in the PLM, PLS, and units. Most of the dug wells in the PLWF and PLWS dry up during the pre-monsoon period; hence bore wells and dug cum bore wells are suggested for these areas. Drainage of the study area exhibits a dendritic to the sub-dendritic pattern because of the presence of more or less homogeneous lithology. The present study shows that the area has plain land with a gentle slope, which is responsible for infiltration and groundwater recharge area. Based on drainage and ground slope studies the possible suitable sites of 78 water conservation structures were proposed in the PG-4 watershed.

Acknowledgements The author Sandip K. Sirsat is acknowledging the Chhatrapati Shahu Maharaj Research Training and Human Development Institute (SARTHI), Pune, is an autonomous Institute of the Government of Maharashtra for financial assistance, in the form of CSMNRF-2019 fellowship is awarded to one Author (Sandip K. Sirsat). The authors also thank the Director, Government Institute of Science, Aurangabad, Maharashtra for their support in creating laboratory facilities which helped in carrying out this work.

References

Anuradha CT, Prabhavathy S (2010) Water resources management for Virudhunagar District using remote sensing and GIS. Int J Earth Sci Eng 3(1):55–61

Abiola O, Enikanselu PA, Oladapo MI (2009) Groundwater potential and aquifer protective capacity of overburden units in Ado-Ekiti, southwestern Nigeria. Int J Phys Sci 4(3):120–132

Babar M (2005) Hydrogeomorphology: Fundamentals, Applications and Techniques. New India Publishing Agency, New Delhi

Central Ground Water Board Report of Buldhana District, Central Region Nagpur (2019).

Chandramohan R, Kanchanabhan TE, Vignesh NS (2019) Identification of artificial recharges structures using remote sensing and GIS for arid and semi-arid areas. Nat Environ Pollut Technolo 18:183–189

Choudhari PP, Nigam GK, Singh SK, Thakur S (2018) Morphometric based prioritization of watershed for groundwater potential of Mula river basin, Maharashtra, India. Geology, Ecology, and Landscapes 2(4):256–267

Das S, Behera SC, Kar A, Narendra P, Guha S (1997) Hydrogeomorphological mapping in groundwater exploration using remotely sensed data—A case study in Keonjhar district, Orissa. J Indian Soc Remote Sens 25:247–259

Deshpande GG (2012) Geology of Maharashtra, published by the Geological Society of India, Bangalore, pp 156–157

ESR (2016) Environmental status report of Nanded city (2014–15). Report prepared by TERI India, pp 12–16.

Gale I, Dillon, P (2006) Strategies for managed aquifer recharge (MAR) in semi-arid areas. United Nations Environment Programme, Division of Technology, Industry and Economics, UNESCO IHP, pp 1–30

Gurav C, Babar M (2019) Hydrogeomorphological studies for groundwater potential of Mangyal nala watershed of Lendi River, Nanded district, Maharashtra. J Geosci Res 2:61–66

Hancock P, Hunt R, Boulton A (2009) Hydrogeoecology, the interdisciplinary study of groundwater dependent ecosystems. Hydrogeol J 17(1):1–3

Jadhav SB, Kalbande SR, Vikhe SD (2009) Evaluation of cement nala plug ARPN. J Engg & Appl Sci 4(1):88–91

Jaiswal RK, Mukherjee S, Krishnamurthy J, Saxena R (2003) Role of remote sensing and GIS techniques for generation of groundwater prospect zones towards rural development—an approach. Int J Remote Sens 24(5):993–1008

Khanday MY, Javed A (2017) Hydrological investigation in the semi-arid Makhawan watershed, using morphometry. Appl Water Sci 7:3919–3936

Kudrna K, Sindelaoova M (2006) Principles of hydrogeomorphology as a basic precondition for solution of territorial structure of unitary system of agricultural, forest and water management. J Central Eur Agric 7(4):669–676

Kumar U, Kumar B (2010) Ground water targeting in hard rock terrain using remote sensing techniques in Sanjai river watershed, Jharkhand, Abs. regional workshop on exploration, Development and management of ground water in hard rocks with special reference to Jharkhand state

Lokesh N, Gopalakrishna GS, Mahesh MJ (2007) Hydrogeomorphological studies in Kallambella Watershed, Tumkur district, Karnataka State, India using remote sensing techniques and GIS. J Indian Soc Remote Sens 35(1):97–105

Loague K, Heppner CS, Mirus BB, Ebel B, Ran Q, Carr A, BeVille SH, Vander K (2006) Physics-based hydrologic-response simulation: foundation for hydroecology and hydrogeomorphology. Hydrol Process 20(5):1231–1237

Madhnure P, Tambe J (2012) Assessment of groundwater pollution in the vicinity of MIDC industrial complex, Nanded district, Maharashtra, India. Mem Indian Soc App Geochem 265–278

Madhnure P (2014) Groundwater exploration and drilling problems encountered in basaltic and granitic terrain of Nanded district, Maharashtra. J Geol Soc India 84:341–351

Murthy KS (2000) Groundwater potential in a semi-arid region of Andhra Pradesh: a geographical information system approach. Int J Remote Sens 21(9):1867–1884

Muthamilselvan A (2017) Delineation of ground water potential zone in Sweta sub-watershed using remote sensing and GIS in parts of Perambalur district of Southern India. J Geomat 11(1):25–30

Pallard B, Castellarin A, Montanari A (2009) A look at the links between drainage density and food statistics. Hydrol Earth Syst Sci 13(7):1019

Patil SG, Mohite NM (2014) Identification of groundwater recharge potential zones for a watershed using remote sensing and GIS. Int J Geomat Geosci 4:485–498

Rajaveni SP, Brindha K, Elango L (2017) Geological and geomorphological controls on groundwater occurrence in a hard rock region. Appl Water Sci 7:1377–1389

Ramaiah SN, Gopalakrishna GS, Vittala SS, KMD, Najeeb (2012) Geomorphological mapping for identification of ground water potential zones in hard rock areas using geo-spatial information—a case study in Malur Taluk, Kolar District, Karnataka, India. Nat Environ Pollut Technol 11(3):369–376

Rani VR, Pandalai HS, Sajinkumar KS, Pradeepkumar AP (2015) Geomorphology and its implication in urban groundwater environment: case study from Mumbai, India. Appl Water Sci 5:137–151

Ravindran KV, Jeyaram A (1997) Groundwater prospects of Shahba Tehsil, Baran district, eastern Rajasthan, A remote sensing approach. J Indian Soc Remote Sens 25(4):239–246

Scientists of Geosciences Group RSA-A, NRSC (2015) Groundwater manual, concepts of ground water prospects mapping for Rajiv Gandhi National Drinking Water Mission Project, Ministry of drinking water and sanitation, National remote sensing centre (NRSC) and Indian space research organization (ISRO) Department of Space, Hyderabad

Selvakumar S, Chandrasekar N, Kumar G (2017) Hydrogeochemical characteristics and groundwater contamination in the rapid urban development areas of Coimbatore, India. Water Resources and Industry 17:26–33

Senthilkumar M, Gnanasundar D, Arumugam R (2019) Identifying groundwater recharge zones using remote sensing & GIS techniques in Amaravathi aquifer system, Tamil Nadu, South India. Sustain Environ Res 29:15. https://doi.org/10.1186/s42834-019-0014-7

Sharma A, Shukla J (2015) A remote sensing and GIS based approach to evaluate the ground water prospects of Beghain watershed, Panna and Satna districts of M.P., India: a case study. J Geol Soc India 86:733–741

Shimpi S, Mahajan C, Rokade VM (2014) Drainage morphometric approach for evaluation of water conservation potential and artificial recharge structures site suitability. Gondwana Geol Mag 14:57–64

Shinde S, Pandurang, CP, Popatkar B, Choudhari N (2020) Assessment of groundwater quality using GIS in Thane Municipal Corporation, Maharashtra, India. Model Earth Syst Environ 7:1739–1751

Siebert S, Burke J, Faures JM, Frenken K, Hoogeveen J, Döll P, Portmann FT (2010) Groundwater use for irrigation—a global inventory. Hydrol Earth Syst Sci 14:1863–1880

Sidle RC, Onda Y (2004) Hydrogeomorphology: overview of an emerging science. Hydrol Process 18(4):597–602

Singh P, Thakur JK, Kumar S (2013) Delineating groundwater potential zones in a hard-rock terrain using geospatial tools. Hydrol Sci J 58(1):213–223

Surrette M, Allen D, Journeay M (2008) Regional evaluation of hydraulic properties in variably fractured rock using a hydro structural domain approach. Hydrogeol J 16(1):11–30

Todd DK (1980) Groundwater hydrology, II. John Wiley & Sons, New York

Varade AM, Khare YD, Yadav P, Doad AP, Das S, Kanetkar M, Golekar RB (2018) Lineaments the potential groundwater zones in hard rock area: a case study of basaltic terrain of WGKKC-2 watershed from Kalmeswar Tehsil of Nagpur District, Central India. Journal of the Indian Society of Remote Sensing 46(4):539–549

Zaveri E, Danielle S, Fisher-Vanden K, Frolking S, Lammers RB, Wrenn DH, Prusevich A, Nicholas RE (2016) Invisible water, visible impact: groundwater use and Indian agriculture under climate change. Environ Res Lett 11(8):084005. https://doi.org/10.1088/1748-9326/11/8/084005

Chapter 7
Suitability of Groundwater for Irrigation: A Spatial Analysis with Reference to Rohtak District

Vinod Kumar and Neeraj

Abstract The rapid increase in population and urbanization has pushed farmers to grow on even marginal land available to them and at the same time, in the light of increasing population pressure on land, land and water resources are being used even more regressively. Resultantly as compared to the past indiscriminate and excessive use of agricultural land has led to the degradation of all land-based resources including water. It is a well-known fact that the prosperity of agriculture in the area is the outcome of many factors. It is widely accepted that after the green revolution, irrigation has emerged as the single most crucial factor determining agricultural changes and the intensity of use of land. The intensity of use of green revolution technology is largely governed by assured irrigation facilities, as most of the areas that have less irrigational potential or have inferior quality of irrigation water lagged in terms of agricultural performance. Haryana lies in the heart of the green revolution belt, and its prosperity of agriculture is based on 'canal-tube-well' irrigational sources. Largely, agriculture in northern and southern districts of Haryana is tube-well based, whereas, central and western districts are dependent on canal irrigation. Given the richness of available water resources in Haryana, it emerged as the powerhouse of agricultural growth in the country. Off late, unwisely use of water resources has raised many issues for sustainable agriculture. Now, many parts of the state are facing problems with falling water tables or rising water tables, due to indiscriminate use of groundwater. Given the dependence on groundwater, quality is also a very crucial aspect in the case of Haryana in general and Rohtak in particular. As many parts of the state are struggling due to the contamination of groundwater with arsenic, mercury, and other heavy metals. Taking a clue from it, in this paper, an attempt has been made to analyze the quality aspect of groundwater using sample data collected by the Central Ground Water Board (CGWB). Using different parameters on the quality

V. Kumar
Geography in Sakshi Malik Government College for Girls, Mokhra, Rohtak 124001, Haryana, India
e-mail: vinodkumar7262@gmail.com

Neeraj (✉)
Department of Geography, Baba Mastnath University, Rohtak 124001, Haryana, India
e-mail: nirajsehrawat0@gmail.com

A. Dubey et al. (eds.), *Promoting Sustainability Through Water Management and Climate Change Adaptation*, Advances in Geographical and Environmental Sciences,
https://doi.org/10.1007/978-981-99-5479-7_7

of groundwater (i.e. Electrical Conductivity, pH, TDS, etc.) suitability for its use for agricultural purposes was analyzed, and found that groundwater quality is not uniform all over in the district of Rohtak. Urgent attention is required in this regard to ameliorate problems about the use of groundwater for irrigational purposes.

Keywords Water quality for irrigation · Suitability · Green revolution technology · TDS · EC · pH

7.1 Introduction

Water has emerged as one of the most essential inputs for the development of agriculture ever since man has realized its importance for agricultural growth and prosperity. More specifically, with the onset of the green revolution and the adoption of newer technology agriculture has transformed into even more water-intensive agriculture. Sowing of crops with higher water requirements under assured irrigation facilities has made agriculture more intense water using economic activity. In this scenario, all the governments are pushing for rapid expansion of irrigational facilities to newer areas. It is found in the review of literature that most of the time, governments' policy thrusts on the expansion of acreage under irrigation, ignoring qualitative aspects of water that is available for irrigation. Scientifically, water is a universal solvent, having the ability to dissolve many substances. Normal water changes its chemical properties under the influence of different elements; therefore, suitability of water for irrigation is ascertained based on its chemical properties and changes in it.

Evaluation of the quality of water for irrigation is equally important, especially in the context of arid and semi-arid regions as these areas are always in need of assured irrigation to sustain. Determination of the quality of water for irrigation purposes is also significantly important as some crops do not have a very high salt tolerance capacity. And these kinds of crops show a fall in the quality and quantity of produce due to poor water quality.

As agriculture is the mainstay of millions of people across the world, every governmental policy betterment and welfare of farmers is the main policy thrust. The development of agriculture has provided stability with an increased carrying capacity that in turn resulted in a further increase in population. The path of food sufficiency in any society and agricultural prosperity has not remained very simple.

In the case of Indian agriculture in the pre-green revolution period agriculture was mainly sustained at the mercy of natural forces. The contours of agricultural development were largely determined by natural forces in general and more specifically by the quantum of rainfall in any area. Quantity of agricultural production and quality (yield) both were synchronized with rainfall amount, timing, and duration. During the lean year, drought was the main determinant of low produce and resultant famine. This period had many deaths in the absence of sufficient food. In similar ways, the flood also plays a very crucial role in the success and failure of agriculture in an area.

Frequent failure of agriculture to meet the needs of the country made agri-scientists look for newer solutions.

The new solution was the change in agricultural technology, it is widely known as package technology. This constitutes the use of HYV seeds along with assured irrigation, fertilizers, and pesticides. In the context of this paper, it must be noted that assured irrigation provided the base for the green revolution. With the assured irrigation, agriculture witnessed qualitative as well as quantitative changes as it led to changes in production and yield. Given the importance of irrigation economists, geographers, and agri-scientist have always been concerned with its expansion. Expansion of irrigation is seen as the most potent aspect of the spread of agriculture and its development. It is often found that most of the time studies related to irrigation have concerned themselves with the expansion of irrigation in terms of the quantity of area under irrigation, ignoring the quality of water for irrigation.

Given the importance of water quality, scientists have identified many methods to quantify and determine water quality using many indices. Some of the indices are **Sodium Adsorption Ratio (SAR), Residual Sodium Carbonate (RSC), Magnesium Hazard (MH), Permeability Index, and Piper diagram or trilinear diagram** have been widely used in scientific advent to determine water quality for irrigational purposes. In the present paper above discussed indices have been used to ascertain irrigation water quality in the Rohtak district of Haryana.

7.2 Study Area

The present study is based on the Rohtak district of Haryana. Rohtak lies between 28°96′ North latitude and 76°29′ East longitude and it is well known for its agricultural and industrial prosperity. Rohtak is blessed with fertile alluvial soil and it has good climatic conditions for prosperous agriculture. With the spread of the green revolution in this central part of Haryana, Rohtak has witnessed a sea change in cropping patterns from millets to finer water-intensive crops (i.e. Rice). Rohtak is very conducive for irrigation, primarily due to its saucer-like geology and because of the availability of canals and tubewells as irrigation resources. With the increase in urbanization and growing population future population sufficiency would be based on the judicious utilization of available water resources. And information regarding water quality and changes in it would be very handy for policymakers to work toward ameliorating conditions in this regard.

7.3 Meaning and Measurement of Quality of Irrigation Water

There are many and varied aspects that are covered in the quality aspect of water used for irrigation. The quality of irrigation waters differs in various regions and countries. The concentration and composition of soluble salts in water determine its quality for irrigation (Zaman et al. 2018). For the present paper to have a common frame of reference quality of water used for irrigation denotes the numeric value of water salinity (EC), sodium hazard (sodium adsorption ratio-SAR), residual sodium carbonates (RSC), and ion toxicity and their levels as prescribed by agricultural scientists in Indian conditions. In the present research paper, the quality of irrigation water is asserted using different parameters as discussed in below sections.

7.4 The Objective of the Study

Many field surveys, research papers, and many farmers have reported that even though Rohtak is blessed with favorable agricultural conditions including assured irrigation, some crops in some areas are not responding to the inputs that are provided. Therefore, by the means of this study, an attempt is made to ascertain the quality of irrigation and describe the problem being properly faced by farmers. The main objectives of the present research paper are as follows:

1. To identify the quality status of water available for irrigation in the study area.
2. Identify ideal tolerance limits of different crops concerning.
3. Suggest cropping pattern changes based upon water quality.

7.5 Source of Data and Methodology

The present study is entirely based on secondary data collected from the annual report for the year 2018–2019 of the Central Ground Water Board (CGWB) Ministry of Jal Shakti, Department of Water Resources, River Department of Water Resources, Government of India. At the same time data on cropping patterns is obtained from the Statistical Abstract of Haryana for the period 1966–1967 to 2019–2020. Later on, data is analyzed with the help of Grapher Software and Diagrams Software through various diagrams. The indices and their formulas that have been used in the present study are as follows.

Table 7.1 Sodium adsorption ratio (SAR)

Sodium hazard class	SAR meq/l	Remarks
S1	Less than 10	Ideal
S2	Range between 10 and 18	Good
S3	Range between 18 and 26	Doubtful
S4	Greater than 26	Unsuitable

Table 7.2 Electrical conductivity (μS/cm)

Level	Electrical conductivity (uS/cm)	Type of water	Suitability for irrigation
C1	Range below 250	Minimal saline water	Entirely safe
C2	Between 250 and 750	Moderately saline	Admirable in all conditions
C3	Between 750 and 2250	Toward high salinity	Usable only in permeable soil
C4	And anything above 2250	High to very high salinity levels	Unfair for irrigation

7.5.1 Sodium Adsorption Ratio (SAR)

One of the most popular indices for asserting water quality is sodium adsorption ratio (SAR).

The SAR denotes chemical property of water in a relative ratio of Na^+ ions to Ca^{2+} and Mg^{2+} ions present in the water sample. Its chemical equation is as under:

$$SAR = NA^+/((1/2(Ca^{2+} + Mg^{2+}))1/2),$$

Where Na^+, Ca^{2+}, and Mg^{2+} are all measured in meq/L (milliequivalents per liter). Scientifically and agronomically the SAR range for irrigation water can be best described as **S1, S2, S3, and S4** classes that are discussed in Table 7.1.

7.5.2 Electrical Conductivity (EC)

Most of the crops have different levels of salt tolerance level, and same is measured via electrical conductivity (EC). Higher the level of the conductivity, the greater its salt content in water having differential impact on productivity and performance of crop. In general the electrical conductivity of water for irrigation can be determined and ascertained as μS/cm (micro Siemens/cm) or dS/m (deciSiemens/m) as explained in Table 7.2.

7.5.3 Sodium Percentage (Na%)

In some crops sodium percent (SSP) is used to measure sodium hazards that are measured as sodium percentage (Na %) in sampled water and it is derived using the equation as given below:

$$Na\% = (NA^{+} + K^{+}/Ca^{2+} + Mg^{2+}) + K^{+}) * 100,$$

In India, BIS (1991) standard has set a threshold limit of 60% sodium percentage for irrigation water to get the optimum yield and production. Wilcox (1955) through his diagram known Wilcox diagram shows graphical representtation of the electrical conductivity and sodium percent.

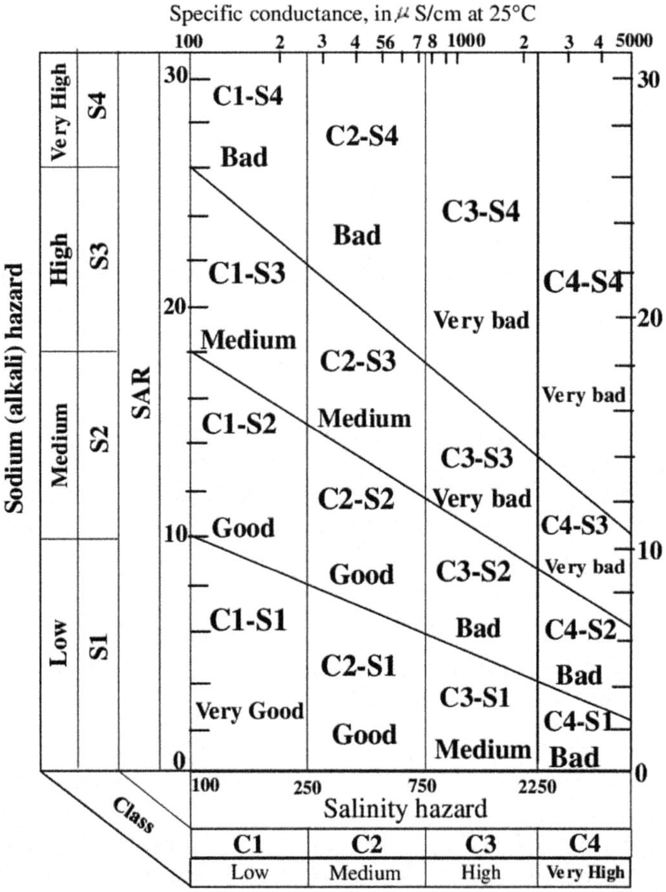

7.5.4 *Magnesium Hazard (MH)*

Magnesium is considered as one of the prime nutrient for healthy growth of plants, and its death can lead to many visible and undesirable results such as it causes yellowing and reduction in the growth and yield of crops. Therefore, for determining the quality of water the concentration of magnesium in water plays a pivotal role. Magnesium hazard was proposed by Szabolcs and Darab (1964) and is calculated as follows:

$$\mathrm{MH} = \mathrm{Mg}^{2+}/(\mathrm{Ca}^{2+} + \mathrm{Mg}^{2+}) * 100$$

Value of magnesium hazard below 50 is regarded as beneficial for most of the crops and it is recommended for use, irrigation water greater than 50 MH is not suitable for irrigation.

7.5.5 *Kelly's Ratio*

In the Kelly's (1940) ratio negative impact of sodium dissolved is measured. And it is calculated using equation given below:

$$\text{Kelly's Ratio}: \ \mathrm{Na}^{+}/(\mathrm{Ca}^{2+} + \mathrm{Mg}^{2+}).$$

In case of value of Kelly's ratio greater than one indicates higher levels of sodium in water. Therefore, water with Kelly's ratio of less than one is suitable for irrigation, and other wise it is regarded unfit for the purpose of irrigation.

7.6 Analysis and Findings

Keeping in mind the objectives of the study analysis of the paper is also subdivided into three coherent parts. Initially, a general account of different water quality for irrigation purposes is presented. In later parts discussion about the ideal tolerance limit of different crops associated with the study area is presented. And at the same time, an attempt has been made to suggest cropping pattern changes taking a clue from the quality of water available for irrigation. Data obtained from CGWB is analyzed using different methods to have a fair understanding of water available for irrigation in terms of its chemistry. It was found that water quality is not uniform as it tends to vary a great deal. For the year 2018–2019, SAR is within an optimum level in all the selected sample sites. Within it, as per the class categorization, nearly 70 samples lie in the excellent category. This shows a healthy picture regarding SAR and its probable impact on agriculture. As against SAR, EC in the selected sample sites seems to be more problematic. As per data, 7 sites have very high EC and it can be termed that in the context of EC more than 60 percent area lies in the unsustainable category. Similarly, in the context of sodium percentage picture is also not very rosy, as out of total samples more than half have unsafe or very high sodium, making it highly unsustainable for agriculture.

Similarly, KR and MH ratios are also presenting a similar problematic picture. In the context of these indices, it is found that more than 80 percent of sample sites have unsuitable water quality levels. In the context of TDS also water quality can be categorized as moderate or hazardous.

Figure 7.1 shows U.S. Salinity Laboratory (USSL) diagram to determine SAR levels in the selected sample size of Rohtak District, 2018–2019, and a careful analysis of the diagram shows that none of the samples in the Rohtak district has an ideal level for cropping based upon SAR and EC. No site lies in areas of good SAR and EC association grids (i.e. S1C1 of S1C2), only Lakhan Majara lies on the boundary of the grid of S1C2 and S1C3 making it a bit more conducive for cultivation. The rest all the other sites have poor EC and SAR associations hence they do not have good quality irrigation water and actions must be taken to ameliorate the water quality status in the region. Similarly, in Fig. 7.1 associations between percentage sodium and EC are depicted. Here too only two sample sites (Lakhan Majara and Bhali Anandpur) have an ideal association between two selected variables. Rest all the other samples have poor or unsuitable status about percentage sodium and EC. Farmers have to alter their cropping pattern so that they can gain maximum present water quality. Here from Table 7.3, an attempt has been made to have made a comparison of Ideal EC levels for different crops sown in the study area with actual levels of EC found for water available for irrigation purposes. Interestingly, it is found that the study area has a largely cropping pattern based on crops such as Wheat-Rice-Barley-Sugarcane, but EC for irrigation water does not support the cropping of sugarcane or rice. The quality of water for irrigation is so poor about EC in some sample sites that the need of the hour is to shift to wheat or barley or any other crop that bears higher EC levels.

7.7 Cropping Pattern and Ground Water Quality in Rohtak

Cropping pattern in an area is the outcome of many factors including agro-climatic conditions, physical, ecological factors, and the quality & suitability of underground water is one such critical factor. The cropping pattern denotes the aerial strength of any crop concerning the gross cropped area. Predominantly, given its geo-climatic condition, Rohtak is blessed with all three cropping seasons namely, Rabi, Kharif, and in some areas Zaid. As shown in Fig. 7.2, in the 1960s Rohtak, wheat as a finer crop, and two millets (Bajara and Jowar) were occupying the main position in the cropping pattern. After 1975–1976 Bajara lost its position to oilseeds with the expansion of irrigation facilities. Up to the too early 1990s, most of the other crops including all millets lost their race to rice and wheat since then both the finer crops have emerged as the main crops in the Rohtak district. With the spread of irrigation,

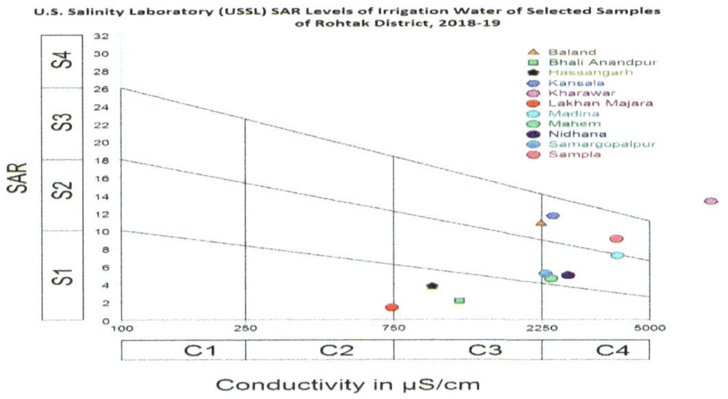

Fig. 7.1 U.S. Salinity Laboratory (USSL) diagram to determine SAR levels in selected sample site of Rohtak District, 2018–2019 (*Source* Calculated by researcher and plotted using software for the sampled villages)

Table 7.3 Water quality indices and their categorization of selected sample sires in Rohtak district

Indices	Range	Class	No. of samples under different classes	Percentage of total Samples
Sodium Adsorption Ratio (SAR)	<10	Excellent	8	72.73
	18–10	Good	3	27.27
EC (μS/cm)	250–750	Good	1	9.09
	750–2250	Doubtful	3	27.27
	>2250	Unsuitable	7	63.64
Percentage Sodium	20–40	Good to Permissible	2	18.18
	40–60	Permissible to Doubtful	3	27.27
	60–80	Doubtful to Unsuitable	6	54.55
Kelly's Ratio	<1	Suitable	2	18.18
	>1	Unsuitable	9	81.82
Magnesium Hazard (MH)	<50	Suitable	1	9.09
	>50	Unsuitable	10	90.91
Total dissolved solids (TDS)	450–2000	Moderate	8	72.73
	>2000	Hazard	3	27.27

Source Categorized by research using data provided by CGWB

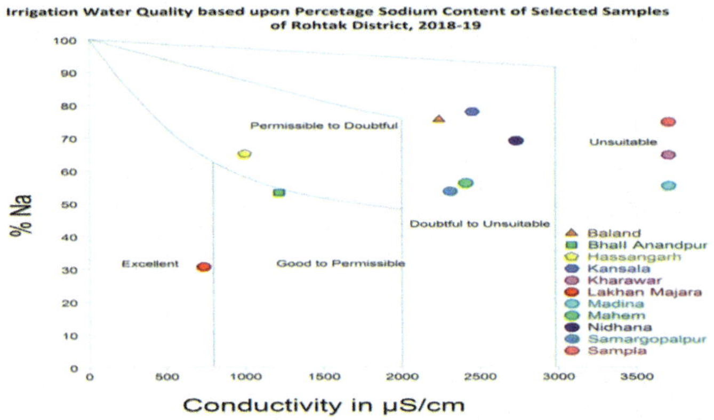

Fig. 7.2 Irrigation water quality-based upon percentage sodium content of selected sample sites of Rohtak District, 2018–2019 (*Source* Calculated by researcher and plotted using software for the sampled villages)

rice gained its prominence in newer areas. Resultantly, wheat and rice occupy nearly 72 percent area of the gross cropped area in Rohtak district.

Data reveals that although there is a very high expansion of irrigation from canals and underground water in the district, still all the regions of the district are not performing equally well in terms of crop production and crop productivity. Studies by Othaman et al. (2020), Lihua Huanga et al. (2017), and Qu et al. (2014) have concluded that one reason for low performance could be attributed to variation in the quality of groundwater in the district and suitability with reference to groundwater for any specific crop.

In this study, an attempt has been made to find out suitable crops regarding groundwater quality in general and more specifically concerning EC of groundwater. This identification of a suitable crop would ameliorate the problem of mismatch between the groundwater quality available and the groundwater quality required by any specific crop Ayers and Westcot (1985). In other word, this identification would help in altering the cropping pattern to the desired effect. Table 7.4 shows the comparison of ideal EC and actual EC levels for different crops in selected sample villages of Rohtak district for the year 2018–2019, it shows that Rohtak has a cropping pattern in which wheat, barley, rice, and sugarcane are prime crops. Surprisingly, Table 7.4 indicates that the groundwater EC level does not support the growth of rice and sugarcane in most of the sampled villages even though these crops have a fairly large share in the Rohtak district. It is even more pathetic than in some of the sampled villages (i.e. Kharawar) groundwater does not support even a single crop from the existing cropping pattern. Given the groundwater EC levels and the growing prominence of groundwater for irrigation, some serious thinking is required in the modification of the existing cropping pattern that supports the present scenario. In that case, suggested crops for Rohtak district can be cotton, fodder crop, increased area under pulses, and oilseeds (Fig. 7.3).

Table 7.4 Comparison of Ideal EC and actual EC levels for different crops in selected sample villages of Rohtak District, 2018–2019

Villages	Actual EC	Ideal EC				Suggested crop
		Wheat	Barley	Rice	Sugarcane	
Baland	2.24	4	5.3	**2***	**1.1***	Wheat and Barley
Bhali Anandpur	1.22	4	5.3	2	**1.1***	Wheat, Rice, and Barley
Hassangarh	1	4	5.3	2	1.1	Wheat, Rice, Sugarcane, and Barley
Kansala	2.45	4	5.3	**2***	**1.1***	Wheat and Barley
Kharawar	7.9	**4***	**5.3***	**2***	**1.1***	No Crop suits from the existing cropping pattern
Lakhan Majara	0.74	4	5.3	2	1.1	Wheat, Rice, Sugarcane, and Barley
Madina	3.93	4	5.3	**2**	**1.1***	Wheat and Barley
Mahem	2.41	4	5.3	**2**	**1.1***	Wheat and Barley
Nidhana	2.73	4	5.3	**2**	**1.1***	Wheat and Barley
Samargopalpur	2.31	4	5.3	**2**	**1.1***	Wheat and Barley
Sampla	3.9	4	5.3	**2**	**1.1***	Wheat and Barley

Source Calculated by the researcher using CGWB data (Note: * indicates EC levels are beyond permissible limits and hence not suitable for crop referred)

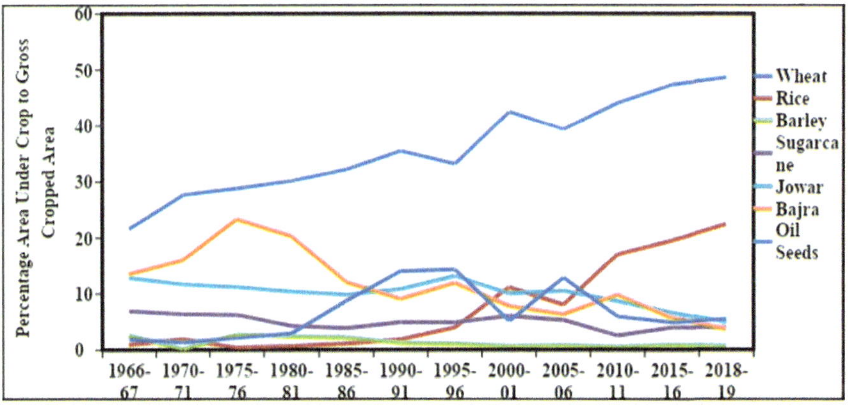

Fig. 7.3 Indicates that the quality of groundwater is not suitable for crops (*Source* Statistical abstract of Haryana for various years)

7.8 Conclusion

In a nutshell, it can be concluded that water is a very vital resource for development in any area and more specifically for agricultural prosperity. Progress and success of the green revolution were also the outcome of assured irrigation. Most of the time, when irrigation is referred it is referred to in the context of the quantity of water available ignoring water quality-related issues. The present paper has revealed that there has been a significant spatial variation in irrigation water quality. Most of the sample sites which were under study failed to get minimum decency level in terms of water quality. Moreover, in most areas existing EC levels of groundwater do not support the growth of rice and sugarcane even though they are grown on large tracks of land. Poor water quality for irrigation not only lowers the production but also causes serious harm to the general productivity levels of the agricultural land. In the context of the area under study need of the hour is to change the cropping pattern in line with the quality of water available for irrigation. In the present context in Rohtak district cotton, fodder crops, pulses, and oilseeds can be seen as substitute crops to match the groundwater EC levels with favorable EC levels required by the crop.

References

Ayers RS, Westcot DW (1985) Water quality for agriculture, FAO Irrigation and Drainage Paper No. 29, Rev. 1, U. N. Food and Agriculture Organization, Rome

Huanga L, Liub X, Wanga Z, Lianga Z, Wanga M, Liua M, Suarez DL (2017) Interactive effects of pH, EC, and nitrogen on yields and nutrient absorption of rice (Oryza sativa L.). Agric Water Manag 194:48–57

Kelly WP (1940) Permissible composition and concentration of irrigated waters. Proc Am Soc Civ Eng 66:607–613

Othaman NNC, Md Isa MN, Ismail RC, Ahmad MI, Hui CK (2020) Factors that affect soil electrical conductivity (EC) based system for the smart farming application. AIP Conference Proceedings

Qu M, Li W, Zhang C (2014) County-scale spatial variability of macronutrient availability ratios in paddy soils. Appl Environ Soil Sci 2014:1–10

Szabolcs I, Darab C (1964) The Influence of irrigation water of high sodium carbonate content of soils. Proceedings of 8th International Congress of ISSS, 2, 803-812

Todd DK, Mays LW (2004) Groundwater hydrology, 3rd edn. John Wiley and Sons, New York

Wilcox LV (1955) Classification and use of the irrigation waters. US Department of Agricultural Technical, Washington, DC

Zaman M, Shahid SA, Heng L (2018) Irrigation water quality. In: Guideline for salinity assessment, mitigation, and adaptation using nuclear and related techniques. Springer, Cham. https://doi.org/10.1007/978-3-319-96190-3_5

Chapter 8
Policy Perspectives and State Responses on Water and Politics in Two States (Telangana State and Andhra Pradesh)

Chalamalla Venkateshwarlu

Abstract Earth's water supply is one of the world's most valuable resources. Water is necessary for both plants and animals to survive. If there was no water there would be no life on Earth. While water is critical to agricultural productivity, it is also essential for animal care and human consumption. Water can be used for a variety of purposes besides simply being ingested. Agriculture relies heavily on water for the development, growth, and production of crops and plants. If there is no water no food for human beings no feed (grass) for animals. Where there is no water (Rain) there are droughts and famines in the World. The role of water in a country's growth and development is undeniable. In countries where irrigation projects have been implemented, food output and economic growth have increased. In the wake of the Krishna water projects, tensions between the two states have erupted. Numerous projects will be built on Krishna by the Telangana state government. Telangana and Andhra Pradesh seem set for yet another collision course over the construction of new irrigation projects on the Krishna River. In view of the above Water and politics in two (Telangana and Andhra Pradesh) states Policy perspectives and state responses are important components in two states. This paper examines the policy perspective of water and politics playing an important role in two states.

Keywords Water recourses · Telangana · Andhra Pradesh · Krishna River · Godavari River · Conflicts

C. Venkateshwarlu (✉)
Department of Political Science, University College of Arts and Social Sciences (UCA&SS), Osmania University, Hyderabad, Telangana State (TS) 500007, India
e-mail: chalamallavenkateshwarlu@osmania.ac.in

A. Dubey et al. (eds.), *Promoting Sustainability Through Water Management and Climate Change Adaptation*, Advances in Geographical and Environmental Sciences,
https://doi.org/10.1007/978-981-99-5479-7_8

8.1 Introduction

Water has always been a communal property resource, and its privatization or even consideration as an economic product would give birth to new tens. For the growth, development, and output of crops and plants, water is needed. Droughts and famines occur around the world where there is a lack of water. Water is important for a country's development and progress. Food production and country development are more prevalent in areas where irrigation projects are underway. Andhra Pradesh and Telangana are heading for a collision path over the Krishna water projects. The Telangana government has chosen to build a number of projects on Krishna. Telangana and Andhra Pradesh appear to be heading for another clash over the development of new irrigation projects on the Krishna River.

8.1.1 Water Conflicts Between Andhra Pradesh and Telangana

Water dispute between Andhra Pradesh and Telangana states of India has always fetched national Attention. With the inter-river water conflict between Andhra Pradesh and Telangana showing no signs of abating, Andhra Pradesh has chosen to file a petition with the Supreme Court, as well as to protest about Telangana's "activities" on its side of the river. Telangana continues its fight for Krishna water rights to be discussed in Parliament. The state administration has resolved to continue its unwavering fight for Telangana's interests on all possible forums in order to secure its rights to share Krishna river water. Interstate disputes over river water are not within the purview of either the Telangana High Court or the Supreme Court, according to that court.

8.1.2 Allocations of Two States

The Andhra Pradesh asserted that the Telangana government is using every means possible to crush the dreams of farmers in the state, including resorting to obvious violations of human rights by forcing the release of massive amounts of water into the sea.

Recent reasons advanced by government officials in the Central Government include the fact that while tribunals and courts have clarified the allocations made to various states, the Supreme Court's directions are required to put them into effect. They also stated that they have witnessed circumstances in which higher riparian states have been indiscriminately draining water and constructing new or extending existing projects and that only a Supreme Court ruling can put things back in order.

8.1.3 Historical Facts and Controversies

Godavari and Krishna are two perennial rivers in peninsular India that flow through both Andhra Pradesh and Telangana, both of which are part of the Deccan Plateau. While 79% of Godavari water is used in Andhra Pradesh and northern Telangana, the remaining 21% is used in coastal Andhra. Similarly, 69% of the Krishna catchment area in Andhra Pradesh is in southern Telangana, 18% in Rayalaseema, and 13% in coastal Andhra. The Krishna River flows approximately 1400 km east from Mahabaleshwar in Maharashtra state to the Bay of Bengal. It runs across Karnataka and is part of the border between Telangana and Andhra Pradesh.

Both Andhra Pradesh and Telangana, which were once one state, posted hundreds of police officers along the Krishna River, which runs through their respective regions. The reason for this is that there is a current conflict over the Krishna waters. Since their separation in 2014, the two states of Andhra Pradesh and Telangana have been at odds. An impromptu decision was taken to divide the water, with the former receiving two-thirds of the vote. While Telangana has since increased its demand to an equal amount, both states are attempting to cancel each other's river projects. They have both claimed that the other is robbing them of their proper portion of the water. Even tension exists along the powerful river's banks.

8.1.4 Aspects of Water Priority: Andhra Pradesh and Telangana

The states of Andhra Pradesh and Telangana must prioritize water and energy efficiency above new projects in order for the two river boards to administer, regulate, operate, and maintain the thirty-six Krishna Basin projects and seventy-one Godavari projects. The agreement is likely to preserve the operations of the Water Resources or Irrigation Departments in the states. The seven-year delay in receiving the letter merely illustrates the tight relationship between the two states over river water sharing. The states have been involved in a war over the use of Krishna water, with Andhra Pradesh proposing a few projects, notably a lift irrigation scheme for Rayalaseema, an area from which the Chief Minister of Andhra Pradesh belongs, and Telangana responding with a half-dozen of its own. Although the Union Ministry recently stated that the Centre will proceed with notifying the boards' jurisdiction, it required nine months to see if a hesitant Telangana government would comply. The seven-year-old state had argued that the notification should follow the completion of a tribunal on Krishna water sharing by the two states, which would broaden the field of reference of the existing Krishna Water Dispute Tribunal II. Telangana had even petitioned the Supreme Court, but the Centre said it would only accept Telangana's plea if it dropped its petition, which Telangana did. Telangana requested that its complaint be submitted to the existing Tribunal to avoid duplication of the probe. The Centre must now ensure that the empowered Boards operate fairly, as the Union government's decision on matters pertaining to the two organization's jurisdiction would be final. As various areas await economic growth, both states have their own reasons for pursuing new water and power projects. Rayalaseema is an arid region, and one of the factors that contributed to the bifurcation was dissatisfaction with the poor exploitation of the two rivers in the then-undivided Andhra Pradesh. Simultaneously, the two countries should prioritize water and energy conservation, as well as enhance the efficiency of irrigation schemes and water reservoirs.

Telangana and Andhra Pradesh must investigate various alternatives and low-cost alternatives. After reviewing the experiences of the restructured Boards, the Centre should consider making the much-touted concept of river basin organizations a reality. The Andhra Pradesh government claims that Telangana is taking Krishna River water for hydel power generation without the consent of the Krishna River Management Board, which is at the Centre of their water conflict. The AP government emphasized that the proposed Rayalaseema Lift Irrigation Scheme was simply intended to use AP's due share of Krishna water and stated that the goal of this endeavor is not to take someone else's water. A farmer is a farmer wherever he or she lives, and water is valuable to everyone. They must live and allow others to live. The two states should work together to guarantee that all farmers have access to water for irrigation and consumption.

The Ministry claimed that their party desired any clashes with other states because the Telangana government generates power at a water level of roughly 800 feet in the Krishna River, and there is nothing wrong with drawing water by lift at 800 feet within their legal share. The Telangana Cabinet sharply condemned the Andhra Pradesh government's illegal construction of the Rayalaseema Lift Irrigation Scheme and the Rajolibanda Right Canal. The Cabinet further chastised the Andhra Pradesh administration for defying the National Green Tribunal's and the Union government's directions in this regard. The Telangana administration has previously petitioned the NGT and filed suits before the Supreme Court against illegal projects in Andhra Pradesh. In addition, the State Cabinet has resolved to take all necessary steps to maintain its rights to Krishna River waters in order to protect the interests of farmers and agriculture in the state. However, the Centre's failure to establish a new tribunal to resolve inter-state conflicts over the Krishna River has proven to be a significant setback for Telangana farmers. Because illegal project construction in Andhra Pradesh causes severe injustice to the irrigation water supply to Palamuru, Nalgonda, Khammam, Warangal, and Rangareddy districts, as well as drinking water to Hyderabad, the Cabinet has decided to take all possible measures to protect farmers' interests and secure Telangana State's fair share of Krishna River waters.

8.2 Krishna Water Disputes Tribunal—River Water Use

The Indian government formed a unified tribunal in 1969 under the provisions of the Interstate River Water Issues Act 1956 to settle river water use disputes between the Krishna and Godavari river basin states. The Krishna River begins in Maharashtra and flows for 300 km through Maharashtra, 500 km through North Karnataka, and 1300 km through Telangana and Andhra Pradesh before flowing into the Bay of Bengal.

8.2.1 Telangana's River Water Utilization Requirements

Krishna River Basin's fourth riparian state is Telangana, which was established just last year. The state wants the federal government to reopen the tribunal because it was not a party to the prior rulings. States in India The tribunal mechanism, according to Karnataka and Maharashtra, is being extended largely to address water issues between Andhra Pradesh and Telangana. This is the second time the states have challenged the tribunal system. Water redistribution will be limited to the states of Telangana and Andhra Pradesh under the enlarged judgement. The central government has finally decided to hold a tribunal hearing on the Krishna River water-sharing problem between the two states, after a lengthy wait.

This final decree governs all imports of water from other rivers into Krishna Basin until riparian governments reach an accord. Recently, Andhra Pradesh's government has begun sending Godavari water to the Krishna delta via the Polavaram right bank canal via the Pattiseema lift. To meet the city of Hyderabad's water supply needs, the Telangana projects Singoor, Manjira, and Yellampalli divert and use Godavari water. A total of 81% of the Godavari water utilized for Hyderabad's needs is regenerated and used for irrigation in Telangana's Krishna basin, as stated in Phase 1 final order's Article VII-A. As a result, the State Cabinet has agreed to build a barrage Jogulamba on the Krishna River at Alampur, inside the adjacent villages of Gummadam, Gondimalla, Velatoor, and Padda Maruru, to channel around 70 TMC of flood water via a pipeline.

As a result, the Cabinet has agreed to address the strategic needs of the Palamuru and Kalwakurthy projects by pumping water into the Yedula Reservoir, which is a component of the Palamuru-Rangareddy Lift Irrigation Scheme. It has also resolved to build the left canal of Pulichinthala and supply irrigation to two lakh acres in the districts of Nalgonda and Suryapet.

Given the availability of river water during the monsoon season, the Cabinet reviewed the feasibility of operating Telangana State's hydropower dams to full capacity. The government is eager to employ water power to lift water from Kaleshwaram and other lift irrigation schemes and fill water bodies, while also lowering the power prices for operating the lift irrigation schemes.

Krishna River resources are derived from legally authorized allotment to the state for irrigation and hydroelectric generation. Telangana is geographically placed above sea level, and there is no possibility to use water for agriculture purposes by gravity, despite the fact that several rivers flow through the state. The government claimed that the state has no choice but to lift water in order to meet its needs. It went on to say that due to the rulers of undivided Andhra Pradesh's neglect over several decades, lands in Telangana have become barren, and farmers in Telangana have been wronged. As a result, the development of irrigation projects in Telangana's new state has become a priority for the state administration, which has developed lift irrigation schemes such as Kaleshwaram to meet Telangana's irrigation needs. Telangana was designated the Rice Bowl of India primarily due to the government's decision to elevate river water to irrigate its farms, and the government plans to build more comparable lift irrigation schemes on both the Krishna and Godavari rivers in the near future.

The government can provide water for both crops by operating hydropower plants at full capacity to raise water from irrigation systems. Apart from utilizing the river waters constitutionally given to Telangana by various tribunals for irrigation, we shall also utilize the water allotted for hydropower generation to avoid the unavoidable lifting of water for irrigation. The State Cabinet has also made a decision in this regard, and the State administration will not consider any opposition. He stated that the administration will battle with anyone and to any length for Telangana's agricultural demands and the welfare of its farmers.

An emphasis has been laid that Telangana State will completely cooperate with its neighbouring States in exercising their appropriate share of the river resources, but that the people of Telangana would not remain silent in the face of the usage of unallocated promised waters. There was also a mention of the then government that had frequently stated both inside and outside the Assembly about the fact that the flood waters would be utilized for Pothireddypadu.

According to the government of Telangana, it is not proper for anyone to disregard the basic natural justice that only when the demands of the River Water Basin are met should out-of-Basin needs be handled. It stated that the demand for hydropower generation has increased as a result of Telangana State's lift irrigation projects, and stated unequivocally that the government will conduct lift irrigation schemes to water Telangana's lands.

It further stated that the Telangana State administration had negotiations with the neighboring Maharashtra State government and built the Kaleshwaram project without generating any problems for the Maharashtra government.

8.2.2 Telangana's River Water Utilization Requirements

Krishna River Basin's fourth riparian state is Telangana, which was established just last year. The state wants the central government to restart the tribunal because it was not a party to the earlier verdicts. The states of Karnataka and Maharashtra oppose the tribunal Krishna Water's Political Perspective. The government proposed that Central Industrial Security Force personnel be stationed at all common reservoirs on the Krishna River shared by Andhra Pradesh and Telangana to better secure and safeguard the two states' interests. The Telangana government agreed to withdraw a complaint filed in the Supreme Court about river water sharing so that the concerns may be resolved by the water dispute tribunals, but he also used this opportunity to reiterate that Andhra Pradesh was doing Telangana an injustice by diverting Krishna waters from the Srisailam reservoir by taking up the Royal scheme. Meanwhile, the government of AP emphasized the impact of a plethora of projects, some without needed regulatory permits, undertaken by Telangana on the Godavari that is damaging to the interests of the lower riparian state.

8.3 Conclusion

Cities in India are depleting peri-urban land and water resources. Informal water merchants transfer water from peri-urban to urban areas in Hyderabad, one of India's fastest-growing cities. This study examines how water tankers are altering water flow patterns and causing issues for peri-urban populations. Beyond the limitations of the state's infrastructure and capabilities, we attribute the growth of informal water markets to power relations between actors. Groundwater depletion and peri-urban water insecurity have both been exacerbated by water transfer. This paper elaborates the water disputes between two states and Central Government interventions to solve the water conflicts at state level.

Chapter 9
Estimation of Surface Runoff Using NRCS CN Method and Geospatial Techniques for Sub-basins Prioritization of Conservation Planning of Ghera Sinhagad Land System, Western Maharashtra

Dnyaneshwar N. Pawar and Sunil Gaikwad

Abstract The study about runoff modeling is necessary to prioritize sustainable land and water resource management. Geospatial techniques have been apply for the grid-based assessment of surface runoff in an engaged region. In this study, surface runoff was predicted using the NRCS SCS Curve Number (CN) method developed by USDA. The study area is Ghera Sinhagad Land System (GSLS), i.e., 131.62 km^2 lies in hilly terrain and the high rainfall zone of the Sahyadri hill ranges of the Deccan plateau, Maharashtra, India. The widespread fieldwork and hydrological measurements were conducted and 361 soil samples were collected for evaluation of soil characteristics. Laboratory work includes the analysis of soil samples, data generation, mapping, an analysis of input parameters and surface runoff estimation using the NRCS (SCS) CN method, database of land use/land cover generated through IRS LISS III FCC image. Grid-wise soil types and LU/LC are used for estmation of HSG. The rainfall grid was prepared using the Isohyetal map of NATMO. Terrain properties were assessed using the DEM. Severe surface runoff (150,000 m^3/year) comprises 3 sub basins including 38.11 km land surface area of GSLS. The surface runoff below 120,000 m /year comprises a 57.33% area of one and two fifth-order sub-basins that are included in the low category of prioritization. Geospatial technologies are potent for predicting surface runoff from inadequate rain-gauged basins for sustainable watershed prioritization conservation and management.

D. N. Pawar (✉)
Department of Geography, KSKW Arts Science and Commerce College CIDCO, Nashik-422 008 Affiliated to S. P. Pune University, M.S., Nashik, India
e-mail: dmauli2007@gmail.com

S. Gaikwad
P.G. Teaching and Research Centre, Department of Geography, S. P. College, Pune 411 030 Affiliated to S. P. Pune University, M.S., Pune, India
e-mail: gaikwad97@gmail.com

A. Dubey et al. (eds.), *Promoting Sustainability Through Water Management and Climate Change Adaptation*, Advances in Geographical and Environmental Sciences,
https://doi.org/10.1007/978-981-99-5479-7_9

Keywords Runoff · Curve Numbers (CN) · NRCS · Geospatial · Prioritization · Conservation

9.1 Introduction

In an Indian province, about 147 Mha of land needs an organized strategy of conservation, Integrated Land Resources Management (ILRM) and implementation (Bhattacharyya et al. 2015). In India, 114.01 Mha the area is degraded, wastelands, and about 23.62 Mha the land is affected by rainfall, and runoff soil erosion (ICAR 2010). Nashik, Ahmednagar, and Pune districts of Maharashtra are extremely affected by water erosion (ICAR 2010). In the severe rainfall region of the Western Ghats, Maharashtra, abundant land facets are degraded through severe runoff-based land erosion and a lack of sustainable conservation strategy (Putty and Prasad 2000). To lighten the undesirable effects of soil erosion, effectual soil conservation plans are requisite at the spatial scale and executive rank. Degraded land facets of Western Ghats.

Geoinformatics technology espetially use of GIS and remote sensing based assessment of surface components of the spatio-temporal scale is essential data for hydrological models (Pietroniro and Prowse 2002; Sun et al. 2010; Rejani 2015; Kumar et al. 2017) and natural resource management studies (Czajkowski and Lawrence 2013; Kumar et al. 2015; Gupta et al. 2016; Tailor and Shrimali 2016; Wang and Xie 2018; Abdulwahd et al. 2020; Sharma and Kanga 2020; Sishah 2021). The geospatial tools facilitate grids-based assessment of surface runoff in the limited or un-gauged catchments. Runoff modeling at the watershed level is essential for sustainable soil and water conservation and management (Nagarajun et al. 2016).

In the Western Ghats region, a high magnitude of rainfall contributes to severe runoff (Putty and Prasad, 2000). The SCS rainfall-runoff models are frequently used for calculating runoff (Gupta et al. 2011). The Curve Number (CN) system has been broadly used to evaluate runoff from the rainfall database (Singh and Mishra 2017). The runoff CN can be effectively incorporated into the experimental models to forecast surface runoff (Tripathi et al. 2002). The SCS CN method of surface runoff rapidly assesses produced runoff in a particular site with sensibly good precision (Gupta et al. 2011). Numerous researchers have used the SCS CN runoff modeling with Geospatial techniques for evaluation and conservation planning (Schultz 1996; Weng 2001; Pietroniro and Prowse 2002; Tripathi et al. 2002; Shivakoti et al. 2011; Gupta et al. 2011; Askar 2013; Vojtek and Vojtekova 2016; Satheeshkumar et al. 2017; Pandey and Stuti 2017; Abraham and Mathew 2018; Anjana and Jinu 2019; Sishah 2021).

The present investigation has been made to examine the quantitative and spatial variation of surface runoff using the NRSC SCS (SCS 1985) Curve number model. The detailed examination of all input parameters and surface runoff variation applied to the systematic prioritization of sub-basins for implementation of a sustainable water and soil conservation programme.

9.2 Study Area

The study area is Ghera Sinhagad Land System (GSLS) lies in hilly terrain and the high rainfall zone of the Sahyadri hill ranges of the Deccan plateau, Maharashtra. It is about 131.62 km^2 and covers 28 villages in the Pune district: Velhe, Haveli, and Bhor tahsils. The geocoordinates of GSLS are 73° 39′ to 73° 51′ E Lon. and 18° 18′ 45″ to 18° 26′ N Lat. It comprises part of the Khadakwasla and Shivganga river catchments. Stream order-wise, 23 sub-basins are delineated for prioritization assessment (Fig. 9.1; Table 9.1).

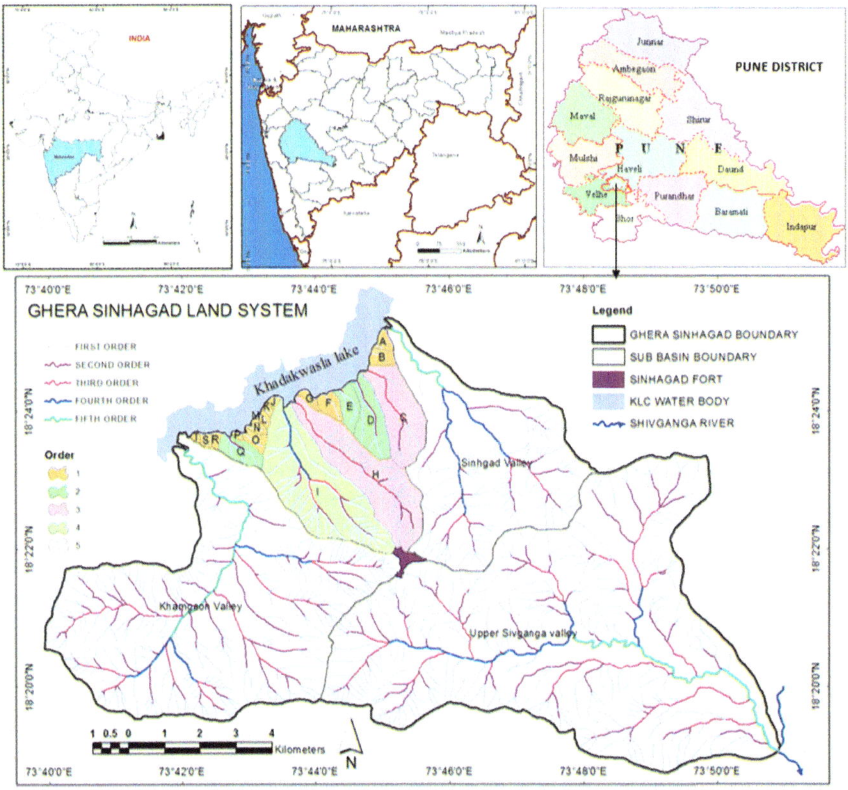

Fig. 9.1 Location map of the study area

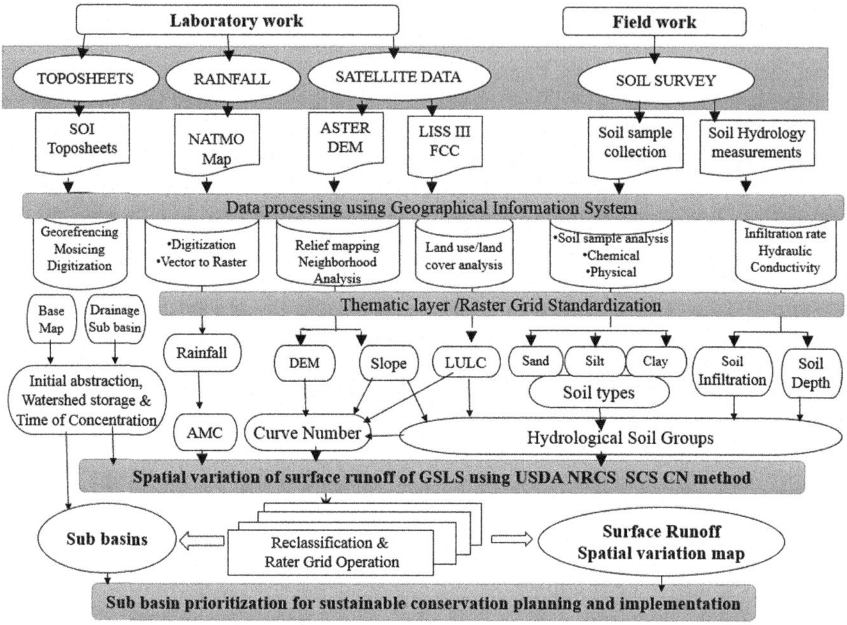

Fig. 9.2 Methodology chart of the present study

9.3 Data and Methods

The present work is classified into laboratory work and field-based activities. Geospatial techniques have been used for precision assessment based on the USDA NRSC SCS-CN method. The detailed methodology is shown in Fig. 9.2. The soil sample collection and hydrological measurement sites were selected based on land use/land cover and slope variation (Fig. 9.2, Table 9.2). A hydrological measurement was including infiltration rate and hydraulic conductivity of soils. The GPS survey was used for land use/land cover and resulted in verification in the study area (Fig. 9.3).

9.4 Estimation of Surface Runoff Using NRCS SCS CN Method

The NRCS-CN method is the most acceptable and used method for estimating surface runoff in ungagged areas. It is a tool for estimating the runoff volume of drainage design, particularly for hill terrain and remote areas. The first step ensures the ratio between actual retention (F) and the maximum potential of basin or watershed storage (S). According to the degree of actual direct runoff volume (Q) to the adequate rainfall (P − Ia). Adequate rainfall estimates using total rainfall subtracted from the initial

Table 9.1 GSLS- area under sub basins order

Sr. no.	Basin order	Total basin	Enclosed area		
			(km^2)	Ha	%
1.	First	14	1.86	186	1.41
2.	Second	3	2.50	250	1.90
3.	Third	2	7.30	730	5.55
4.	Fourth	1	6.31	631	4.79
5.	Fifth	3	113.40	11,340	86.16
Total		23	131.37	13,137	99.81
Sinhagad fort			0.25	25.2	0.19

Table 9.2 GSLS-soil sample collection according to slope variation

Sr. no.	Slope (%)	Slope area (km^2)	Soil sample and field measurement sites
1.	<10	23.37	93
2.	10–20	34.26	134
3.	20–30	21.52	85
4.	30–40	16.06	39
5.	>40	5.4	10

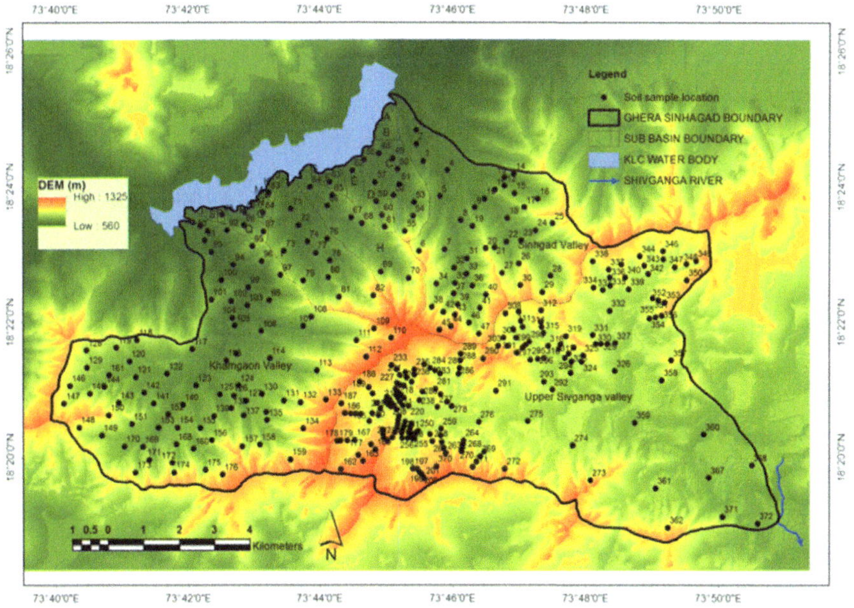

Fig. 9.3 Location of soil samples collections and hydrological measurements

abstractions.

$$F/S = Q/(P - Ia) \tag{9.1}$$

From the continuity principle,

$$F = (P - Ia) - Q \tag{9.2}$$

The NRCS SCS-CN method defined the significance of the initial abstraction (Ia) to be almost equal to 20% of the basin storage (S) The another part holds that,

$$Ia = 0.2S \tag{9.3}$$

Here, Iameans the portion of the rainfall that will not include as runoff. Substituting Eq. (9.5) into Eq. (9.3) and resolving for Q provides the typical appearance of the NRCS-CN process solving the Eqs. 9.1 and 9.2 consecutively,

$$Q = (P - 0.2S)\,2/P + 0.8S \tag{9.4}$$

The third part includes that,

$$S = (25400/CN) - 254 \tag{9.5}$$

Peak discharge (Qp) can be calculated using equation

$$Q = (P - Ia)\,2/(P - Ia) + S \tag{9.6}$$

Curve Number (CN) ranges between 1 and 100. The CN 100 means the entire rain runs off.

Runoff producing rainfall can be deliberate using the equation ($P = 0.7\,R - 21.2$). Here the R means daily rainfall in (mm) or (inch). It is recommended to the hilly topography of Indian conditions.

$$Q = (P - 0.1S)\,2/(P + 0.7S) \tag{9.7}$$

Previous studies in India, regarding runoff estimation have used the following equation

$$Q = (P - 0.2S)\,2/(P + 0.8S) \tag{9.8}$$

The NRCS SCS curve number (CN) is resolute using land use and land cover, the hydrologic soil group (HSG) and the antecedent moisture condition (AMC). HSG comprises four groups, including the soil-infiltration variation acquired for bare soil after continued moistening. The AMC is expressed as three levels of rainfall limits of latent and growing seasons.

9.4.1 Input Parameters of Surface Runoff of NRCS SCS-CN Method

The surface runoff of the GSLS was estimated both spatially and quantitatively using NRCS (SCS) curve number method. The CN was modified to the hilly terrain of Indian conditions. Significant input parameters of runoff have been calculated by applying the field base and secondary database. The analysis and mapping were accomplished using geospatial techniques of GIS, GPS, and remote sensing. The details and results of each parameter have been verified.

9.4.2 Rainfall Variation

The selected study area is an ungauged catchment, and only a single rainfall station exists. The study area's rainfall distribution map was prepared using the NATMO map, and further, it was then processed to prepare a raster grid (Gupta et al. 2016). The rainfall map was converted to a high-resolution raster grid and gave outstanding outcomes. The point data of each grid of the rainfall has been processed for further calculations. The rainfall shows an increasing trend toward the western part of the study area. The rainfall variation shows that the average rainfall of the study area ranges between 900 and 1750 mm (Fig. 9.4). The Khamagaon basin receives the maximum rainfall in the region. Rainfall and runoff have a positive relationship (Dahe and Deshmukh 2018) (Fig. 9.4).

9.4.3 Land Use/Land Cover

Land use/ land cover analysis and mapping of GSLS have been performed using the IRS LISS III FCC of 2003. The area under each class of land use/land cover is delineated, and results show that about 41.18% of GSLS comprises the wasteland category (Fig. 9.5). This land is desperately vulnerable to severe potential surface runoff and water-induced soil erosion. The LULC and its extent is calculated for each grid. It is used to estimate CN and HSG in estimating surface runoff. A higher number of CN is assigned to the land use category, contributing to the highest runoff and vice versa (Savinai 2016; Matej and Vojtekovάb 2019). The LULC accuracy assessment shows a more than 95% level (Table 9.3).

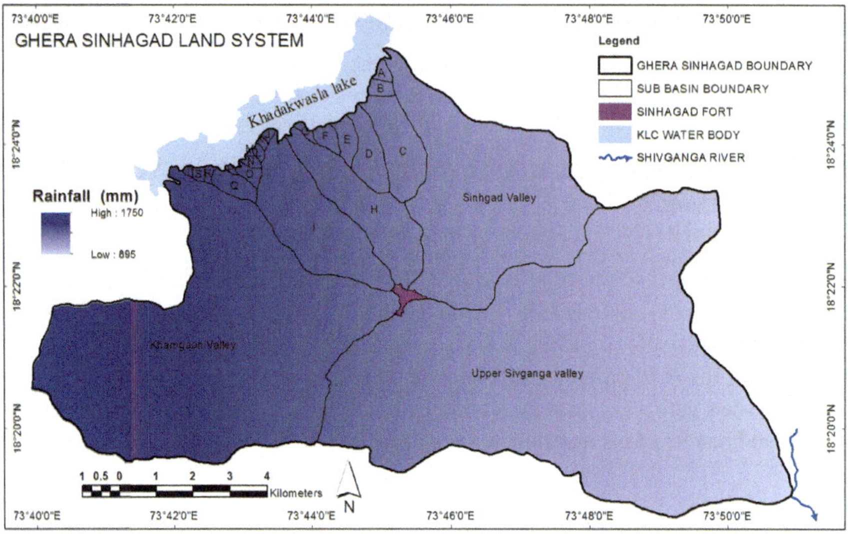

Fig. 9.4 GSLS: rainfall variation

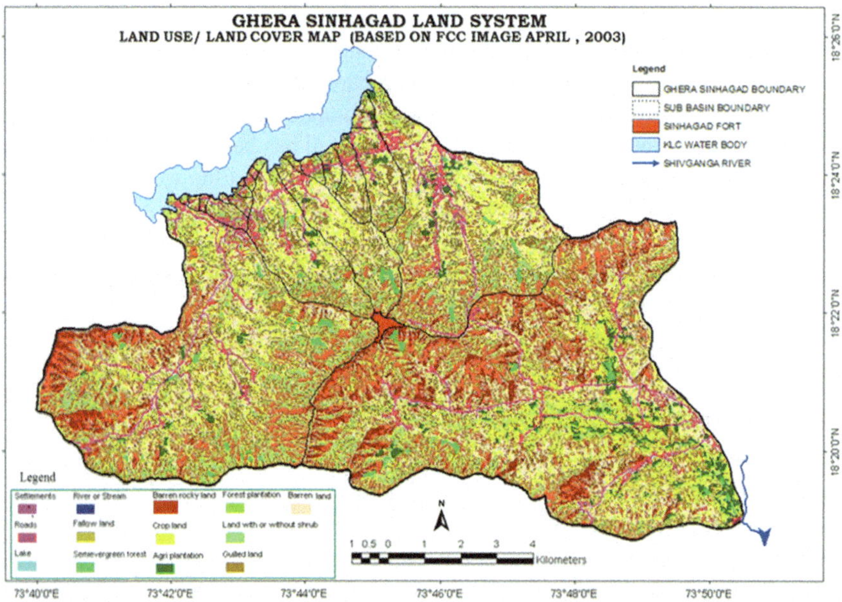

Fig. 9.5 GSLS: land use/land cover (*Source* Pawar 2013)

Table 9.3 GSLS-extent of LULC classes

	1. Built up land			2. Forest land			3. Water body	
Area	Settlements	Roads	Fort	Evergreen forest	Semi evergreen	Forest plantation	Lakes	River
km²	1.42	1.2	0.25	0.79	13.24	2.24	0.01	0.8
%	1.08	0.91	0.19	0.6	10.05	1.7	0.01	0.61
	4. Waste land				5. Agricultural land			
Area	Barren land	Barren rocky	Land with scrub	Gullied/ ravenous	Crop land	Plantation	Fallow land	
km²	17.64	4.7	9.33	22.56	20.42	1.98	3504	
%	13.39	3.57	7.08	17.13	15.5	1.5	2661	

9.4.4 Soil Types

It is estimated using prepared soil type is an indispensable parameter of surface runoff and is required to estimate the hydrologic soil groups. The soil-type map of the GSLS is prepared using soil texture analysis of collected soil samples from the fieldwork. The percentage of sand, silt, clay, and loam is estimated and raster variation maps have been prepared. The soil type map of the GSLS is prepared and the grid-wise soil type's extent is calculated for further runoff analysis methodology (Fig. 9.6; Table 9.4). There is a substantial variation in the soil texture of the GSLS, which has resulted in variation in the soil types, which range from sandy loam (SL), sandy clay loam (SCL), clay loam (CL), loamy sand (LS), Silty clay loam (SiCL), and Silty clay (SiC) (Fig. 9.6; Table 9.4).

9.5 Infiltration Rate Variation

The soil infiltration variation of GSLS was estimated using field measurements. The spatial variation of infiltration of GSLS was prepared based on more the 200 measurement sites using the infiltrometer. Grid-wise database generated for the estimation of CN, HSG, and surface runoff analysis. LULC and slope variation have a notable influence on infiltration variation. The maximum infiltration is recorded in the extensive side slopes valley and agricultural fields. Low infiltration is noticed at hill summit convexities and steep valley-side slopes barren and fallow land, significantly the patches of severe runoff potential (Fig. 9.7; Table 9.5).

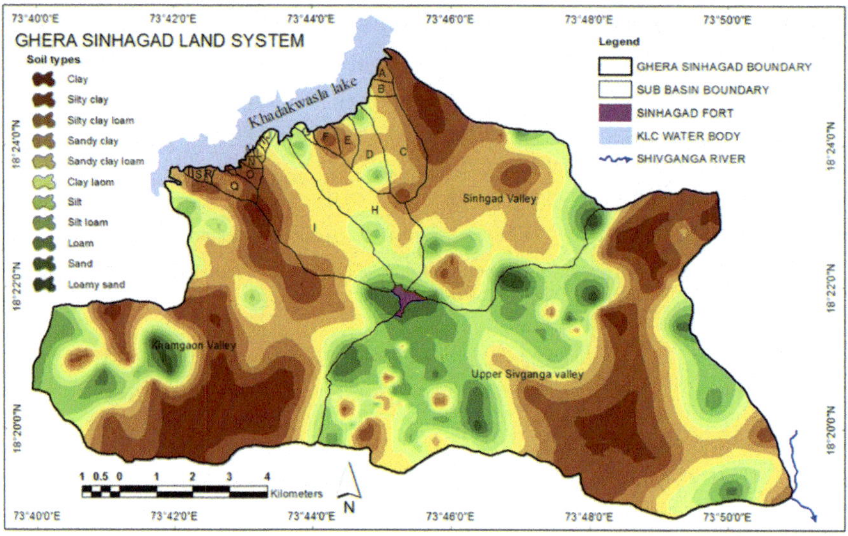

Fig. 9.6 GSLS: Soil types of GSLS

Table 9.4 GSLS-extent of soil types

Area	Clay	Silty clay	Silty clay loam	Sandy clay	Sandy clay loam	Clay loam	Silt	Silt loam	Loam	Sand	Loamy sand
km^2	21.02	19.26	16.93	21.04	12.96	12.42	10.37	11.39	4.69	1.39	0.15
%	15.97	14.63	12.86	15.99	9.85	9.44	7.88	8.66	3.56	1.06	0.12

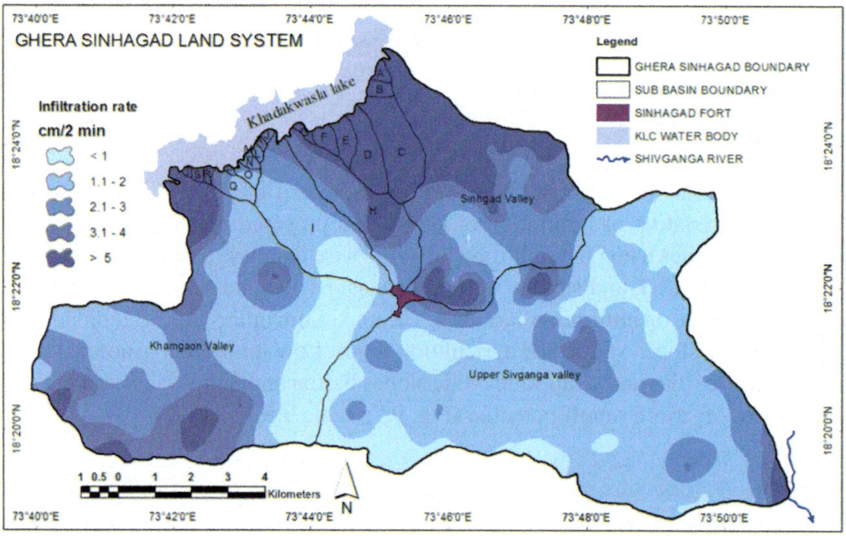

Fig. 9.7 GSLS: infiltration rate variation

Table 9.5 Extent of variation in infiltration rate

Class	Infiltration rate category	Infiltration rate	Area		
		cm/m	km^2	ha	%
1	Very slow	<0.1	1.77	177	1.34
2	Slow	0.1–0.5	2.17	217	1.65
3	Moderate	0.5–2.0	81.7	8170	62.07
4	High	2.0–6.0	43.17	4317	32.80
5	Severe	6.0–12.5	2.81	281	2.13
6	Very severe	>12.5	0.00	0.00	0.00
Total area			131.6	13,162	100

9.6 Hydrological Soil Groups (HSG)

It is essential to calculate the curve numbers of surface runoff modeling. In the present study, about four HSG of A, B, C, and D were classified based on the potential runoff characteristics (Dhruvnarayan 1993). The database used to estimate HSG of GSLS is soil types (Chow et al. 1988; Tailor and Shrimali 2016), soil-infiltration rate, permeability, soil depth, (Fig. 9.8) slope (Fig. 9.9) and LU/LC (Fig. 9.10). Entire input parameters were classified into the same grid and HSG map of GSLS has been prepared (Fig. 9.8; Table 9.6).

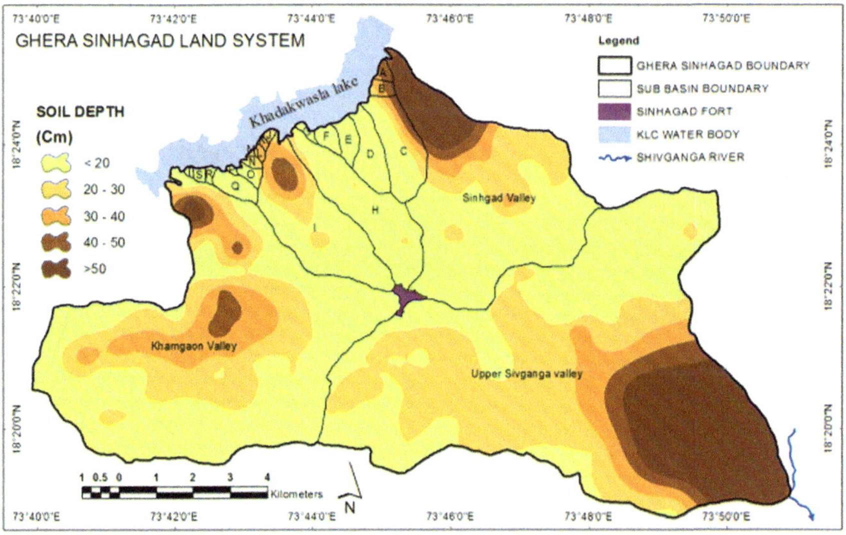

Fig. 9.8 GSLS: soil depth

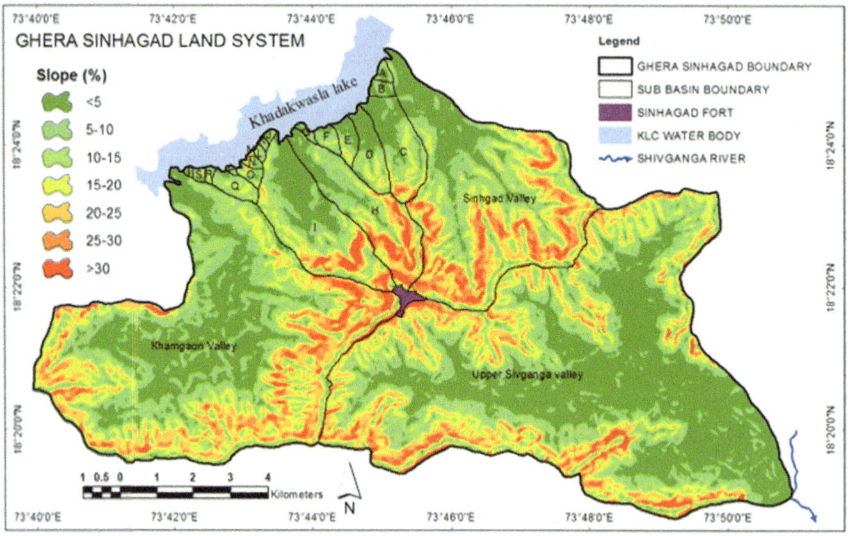

Fig. 9.9 GSLS: percentage slope

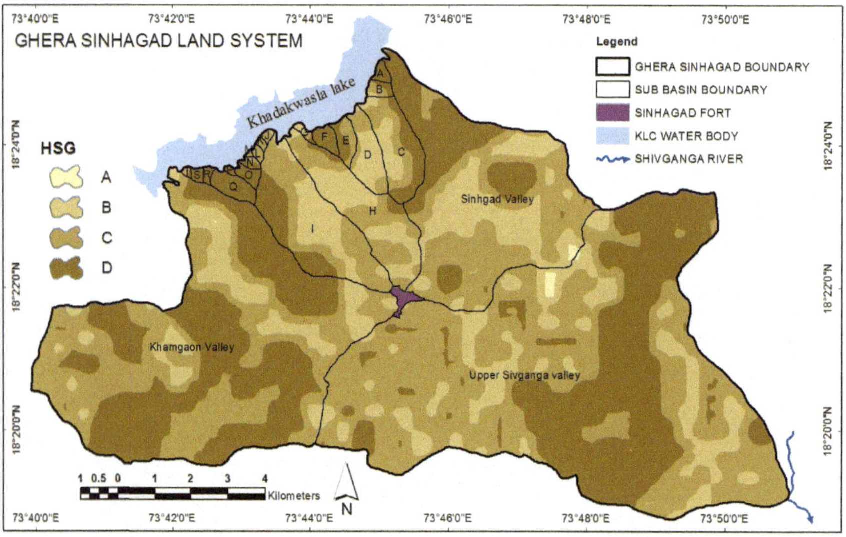

Fig. 9.10 GSLS: hydrological soil groups (HSG)

HSG A: It also characterized the low runoff potential and the high infiltration rate variation in wetted soil. They consist of deep soil, well to excessively drained sandy or gravel texture. Medium slope and agricultural land are included in the group.

Table 9.6 GSLS-sub-basin wise extent of hydrological soil groups (HSG)

Basin order	Sub-basin ID	Hydrological soil groups (Area in km^2)				Total area
		A	B	C	D	
Fifth	GSLS 1	3.042	11.8	17.86	5.3	38.0020
	GSLS 2	6.75	26.08	15.72	8.15	56.7000
	GSLS 3	2.45	10.39	3.73	2.14	18.7100
Fourth	GSLS 12	0.64	4.05	1.19	0.43	6.3100
Third	GSLS 6	0.38	1.19	0.72	0.27	2.5600
	GSLS 11	1.05	3.39	0.31		4.7500
Second	GSLS 7	0.58	0.78			1.3600
	GSLS 8		0.51	0.3	0.02	0.8300
	GSLS 20			0.25	0.37	0.6200
First	GSLS 4			0.13	0.05	0.1800
	GSLS 5	0.03	0.13	0.1		0.2600
	GSLS 9		0	0.17	0.17	0.3400
	GSLS 10	0.06	0		0.03	0.0900
	GSLS 13		0.081			0.0810
	GSLS 14		0.0619			0.0619
	GSLS 15		0.097			0.0970
	GSLS 16		0.0284			0.0284
	GSLS 17		0.023			0.0230
	GSLS 18			0.056	0.2	0.2560
	GSLS 19			0.03	0.023	0.0530
	GSLS 21			0.04	0.06	0.1000
	GSLS 22			0.044	0.7	0.7440
	GSLS 23			0.022	0.034	0.0560
Total		14.982	58.6113	40.672	17.947	131.21

HSG B: The moderate runoff potential, moderate infiltration rate variation, moderately well-drained soils, and moderately fine to moderately coarse textures of soil are characteristics of the group.

HSG C: Soils of high infiltration rates when thoroughly wetted, low infiltration rate, less permeability, steep slopes, and moderately fine to fine soil texture in this group. Fallow land and barren land are comprised under HSG C.Soil.

HSG D: This group indicates soils of very high to severe runoff potential. The soil has very low infiltration rates, clay texture, land with a water table, soil texture with a clay pan, very near the surface, and shallow soils over nearly impervious material.

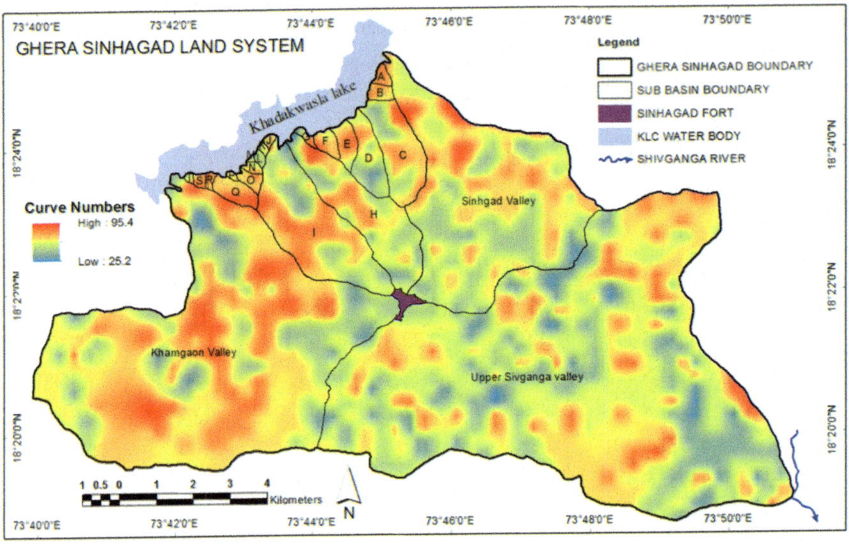

Fig. 9.11 GSLS: Curve Number distribution

9.7 Curve Numbers (CN)

It is a significant parameter of the runoff estimation. The CN has been obtained from USDA NRCS (SCS) method and TSA TR 55 modules. The HSG, LULC, and soil types integrated data deduced to the Indian conditions considered for CN value estimation (Narayana 1993; Vojtek and Vojtekova 2016; Mohamed et al. 2020.). Grid-wise CN was calculated, and the spatial variation map of GSLS has been prepared (Fig. 9.11; Table 9.7).

9.8 Antecedent Moisture Conditions (AMC)

The AMC introduces the moisture content available in the soil at the inception of the rainfall-runoff situation under consideration. AMC governs the initial abstraction and the infiltration rate. The present study has been designed for inferences of average measures of runoff (Table 9.7). The AMC of type II has been considered (Tailor and Shrimali 2016; Abraham and Mathew 2018).

Table 9.7 GSLS-land use /land cover wise values of curve numbers

Sr. no.	Soil types	Soil group	Land use/land cover wise curve numbers						
			Upland paddy	Low land paddy	Fallow land	Settlement	Closed forest	Open forest	Shrubs
1.	Loamy sand	A	95	95	71	77	26	49	33
2.	Sandy loam	B	95	95	80	86	40	69	47
3.	Loamy	C	95	95	85	91	58	79	64
4.	Silty loam	C	95	95	85	91	58	79	64
5.	Clay loam	C	95	95	85	91	58	79	64
6.	Silty clay loam	D	95	95	88	93	61	84	67

9.9 Initial Abstraction, Watershed Storage, and Time of Concentration

Initial abstraction has been calculated using the USDA methodology. The extent of deviation of initial abstraction remarked from 0.76 to greater than 2 inches, and maximum initial abstraction (Ia) was determined to be first-order streams (Table 9.8).

9.10 Results

9.10.1 Spatial Variation of Surface Runoff of GSLS

In the present study, an exercise was attempted for grid-wise estimation of the surface. The *GSLS* has been divided into a 0.003 arc degree (0.105 km^2) scale of 324 m by 324 m. Input parameters of surface runoff computation of the GSLS estimated for the same grid using GIS raster calculator. The spatial runoff variation map of GSLS shows that (Fig. 9.12) severe runoff was recorded to hill summit zones and along the very steep slope. The amount of rainfall received is directly associated with the runoff variation in the area LULC; slope and soil texture play an essential role in determining the volume of direct surface runoff. The area comprised under forest specifies the low potential of surface runoff. The highest potential runoff is noticed in the *Khamgaon* basin and accounts for more than 1.5 lack cubic meter average runoff.

Table 9.8 GSLS-initial abstraction, watershed storage and time of concentration

Sr. no.	Basin order	Sub-basin	Sub-basin (area in ac.)	Hydraulic length(ft)	Average slope (%)	WCN	Tc	Ia	l/w
1.	5	GSLS 1	9389.97	35,685.59	19.30	64.40	4.53	1.11	6.23
2.	5	GSLS 2	14,010.82	54,205.87	15.79	58.34	5.32	1.43	6.75
3.	5	GSLS 3	4620.85	25,633.12	22.81	63.46	2.35	1.15	5.41
4.	4	GSLS 12	1559.23	15,925.15	18.40	63.28	1.80	1.16	4.35
5.	3	GSLS 6	630.12	10,150.89	13.36	72.27	1.16	0.77	3.63
6.	3	GSLS 11	1173.75	17,844.43	24.04	64.53	1.67	1.10	4.11
7.	2	GSLS 7	336.06	7148.93	12.41	62.00	1.19	1.23	3.20
8.	2	GSLS 8	130.72	3467.51	11.07	66.10	0.64	1.03	2.65
9.	2	GSLS 20	151.72	3566.26	16.38	70.50	0.48	0.84	2.73
10.	1	GSLS 4	44.48	577.43	7.60	68.70	0.17	0.91	2.14
11.	1	GSLS 5	63.75	780.84	6.73	55.80	0.33	1.58	2.30
12.	1	GSLS 9	84.02	2457.34	15.84	72.40	0.34	0.76	2.43
13.	1	GSLS 10	31.14	958.00	6.74	63.70	0.31	1.14	1.99
14.	1	GSLS 13	19.92	793.96	13.30	49.00	0.28	2.08	1.82
15.	1	GSLS 14	15.30	718.50	15.23	49.40	0.24	2.05	1.73
16.	1	GSLS 15	23.85	1003.93	14.70	49.90	0.31	2.01	1.89
17.	1	GSLS 16	7.02	396.98	15.06	59.90	0.11	1.34	1.48
18.	1	GSLS 17	30.39	1059.71	15.70	62.00	1.05	1.23	1.98
19.	1	GSLS 18	61.28	1820.86	15.92	64.50	0.34	1.09	2.28
20.	1	GSLS 19	12.97	1043.30	12.50	65.30	0.23	1.06	1.67
21.	1	GSLS 21	23.57	875.98	10.46	58.30	0.27	1.43	1.88
22.	1	GSLS 22	26.93	872.70	12.32	49.20	0.31	2.07	1.93
23.	1	GSLS 23	13.99	688.97	8.64	50.30	0.30	1.98	1.69

9.10.2 Sub-basin Wise Surface Runoff of GSLS

Sub-basin wise surface runoff and area under runoff category of the GSLS calculated for the sub-basin prioritization.

Low runoff (less than 100,000 m^3/year): This class covers 16.53 km^2 (1653 ha 12.56%) land surface of *GSLS*. Sub-basin 2 and 3 comprises almost 12.07 km^2 and 3.98 km^2, respectively. The sub-basin 6 and 8 noticed 0.32 km^2 and 0.15 km^2, accordingly. It can be seen from the map that the maximum area under this class comprises the eastern part of the study area (Table 9.8).

Medium runoff (100,000 m^3 to 115,000 m^2/year):-This category comprises a maximum land surface of 45.43 km^2 (34.51%) of the GSLS. *Shivganga* valley (35 km^2), GSLS 3 (8.13 km^2) comprises under this category. Sub-basin GSLS 6 and 11 covers 1 km^2 area (Table 9.8).

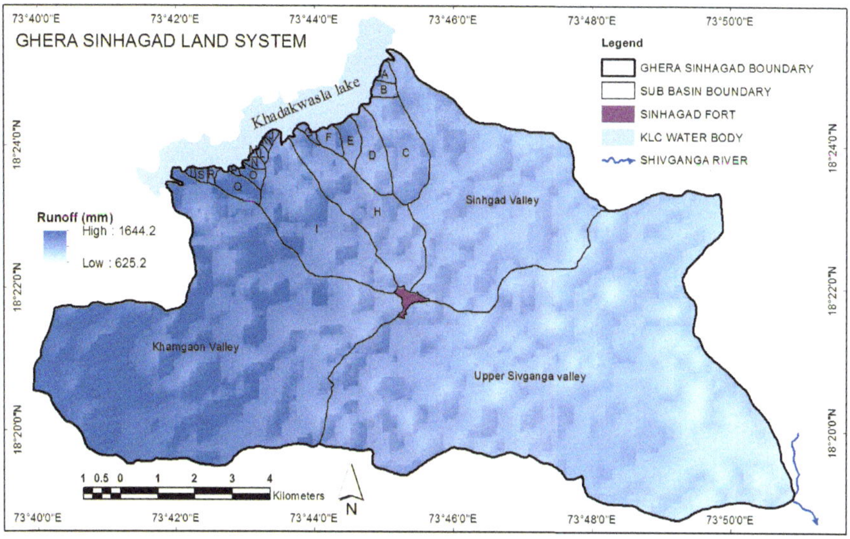

Fig. 9.12 GSLS: spatial distribution of surface runoff

High runoff (115,000 m^3 to 130,000 m^3/year runoff): It comprised 16.11 km^2 (12.24%) of the GSLS and was noticed on bare hill slopes pediment surfaces.

GSLS 3 covers a maximum area (6.58 km^2). Another sub-basin accounts consequently 4 km^2 GSLS 2, 1.51 km^2 of GSLS 11, 1.37 km^2 in GSLS 6; 1 km^2 in the GSLS 1. The GSLS 4, 5, 7, 8, 9, 12 and 14 covers a few patches of this class (Table 9.8).

Very high runoff (130,000 m^3 to 145,000 m^3/year): Total area under this category accounts for 17.65 km^2 (13.41%). It comprises GSLS 1 and GSLS 3 and accounts for 9.65 km^2. Sub-basin GSLS 12 and GSLS 11 cover 4.56 km^2 and 1.70 km^2, respectively (Table 9.8).

Severe runoff (145,000 m^3 to 160,000 m^3/year): High WCN indicates high runoff potential with dominant clay loam texture of the soils. Severe runoff comprises 22.79 km^2 (17.31%) of the total study area. GSLS 1 and 2 account for 19.20 km^2 and 1.50 km^2 area under severe runoff.

Very severe runoff (>160,000 m^3/year): The barren and barren rocky or scrublands are characterized under this class. Very steep slopes and low infiltration hill fringe surfaces contribute to severe runoff (Table 9.8).

It can be observed from Fig. 9.13 (Table 9.8) runoff of more than 160,000 m^3/year is noticed to the western part and consists of a 12.38 km^2 (09.41%) area. GSLS 1 (10.28 km^2) and GSLS 2 (2.10 km^2) land also need an urgent conservation plan (Table 9.9).

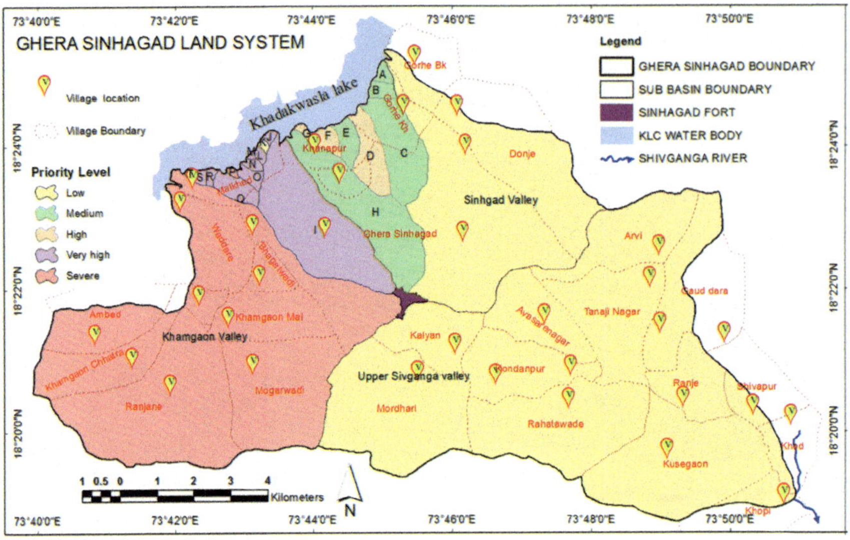

Fig. 9.13 GSLS: sub-basin-wise priority level for sustainable conservation planning

9.11 Prioritization of Sub-basins

The *GSLS* is divided into total 23 sub-basins based on their stream orders. It includes 14 first-order, 3s-order, 2 third-order, 1 fourth-order, and 3 fifth-order basins. Based on runoff assessment of GSLS, the sub-basin-wise priority has been suggested for the conservation planning (Fig. 9.13; Table 9.10).

Table 9.9 GSLS-sub-basin-wise aerial extent and average runoff variation

Sr. no.	Sub-basin	Runoff category (m^3)						Total area (km^3)	Average runoff (m^3)
		Low	Medium	High	Very high	Severe	Very severe		
		<100,000	100,000–115,000	115,000–130,000	130,000–145,000	145,000–160,000	>160,000		
1.	GSLS 1	0.00	0.00	1.00	7.52	19.20	10.28	38.00	152,229.28
2.	GSLS 2	12.07	35.00	4.00	2.13	1.50	2.10	56.80	119,504.54
3.	GSLS 3	3.98	8.13	6.58	0.00	0.00	0.00	18.70	107,592.48
4.	GSLS 4	0.00	0.00	0.18	0.00	0.00	0.00	0.18	123,209.03
5.	GSLS 5	0.00	0.00	0.26	0.00	0.00	0.00	0.26	121,190.06
6.	GSLS 6	0.32	0.87	1.37	0.00	0.00	0.00	2.55	123,501.00
7.	GSLS 7	0.00	0.54	0.82	0.01	0.00	0.00	1.36	135,309.00
8.	GSLS 8	0.00	0.11	0.22	0.21	0.00	0.00	0.53	125,600.40
9.	GSLS 9	0.00	0.00	0.03	0.31	0.00	0.00	0.34	133,857.77
10.	GSLS 10	0.00	0.00	0.00	0.13	0.00	0.00	0.13	136,831.46
11.	GSLS 11	0.16	0.78	1.51	1.70	0.00	0.00	4.15	124,413.57
12.	GSLS 12	0.00	0.00	0.07	4.56	1.67	0.00	6.31	141,954.79
13.	GSLS 13	0.00	0.00	0.00	0.08	0.00	0.00	0.08	143,840.00
14.	GSLS 14	0.00	0.00	0.06	0.00	0.00	0.00	0.06	115,740.86
15.	GSLS 15	0.00	0.00	0.01	0.06	0.03	0.00	0.10	147,693.45
16.	GSLS 16	0.00	0.00	0.00	0.03	0.00	0.00	0.03	148,812.09
17.	GSLS 17	0.00	0.00	0.00	0.12	0.00	0.00	0.12	147,678.54
18.	GSLS 18	0.00	0.00	0.00	0.25	0.00	0.00	0.25	149,264.30
19.	GSLS 19	0.00	0.00	0.00	0.05	0.00	0.00	0.05	150,094.69
20.	GSLS 20	0.00	0.00	0.00	0.22	0.39	0.00	0.61	144,117.53

(continued)

Table 9.9 (continued)

Sr. no.	Sub-basin	Runoff category (m^3)						Total area (km^3)	Average runoff (m^3)
		Low	Medium	High	Very high	Severe	Very severe		
		<100,000	100,000–115,000	115,000–130,000	130,000–145,000	145,000–160,000	>160,000		
21.	GSLS 21	0.00	0.00	0.00	0.10	0.00	0.00	0.10	149,552.08
22.	GSLS 22	0.00	0.00	0.00	0.11	0.00	0.00	0.11	147,246.68
23.	GSLS 23	0.00	0.00	0.00	0.06	0.00	0.00	0.06	151,004.80
Total		16.53	45.43	16.11	17.65	22.79	12.38	131.37	136,532.10

Table 9.10 GSLS-sub-basin priority for conservation planning

Priority level	Priority class	Sub-basin average annual runoff (m^3)	Sub-basin order	Sub-basin Id	Total sub-basin	Total area (km^2)
1	Severe	Above 150,000	1 2	GSLS 19, 23 GSLS 1	2 1	38.11
2	Very high	140,000–150,000	1 2 4	GSLS 13, 15, 16, 17, 18, 21, 22 GSLS 20 GSLS 12	7 1 1	7.70
3	High	130,000–140,000	1 2	GSLS 9, 10 GSLS 7	2 1	1.83
4	Medium	120,000–130,000	1 2 3	GSLS 4, 5 GSLS 8 GSLS 6, 11	2 1 2	8.27
5	Low	Below 120,000	1 5	GSLS 14 GSLS 2, 3	1 2	75.46

9.12 Conclusion

The consequence of this study can apply to sustainable conservation and management of land and water resources in hilly terrain areas. Geospatial techniques have been trustworthy to be an effective tool for estimating spatial runoff. Data analysis, processing, raster tools, satellite images, GPS survey for sampling and validation of results, mapping, and statistical reports with adequate accuracy can be operational through Geospatial techniques. The NRCS SCS-CN model is the most extensively accepted model for estimating surface runoff. The study shows that rainfall and slope factors directly affect surface runoff. The suitable Antecedent Moisture Condition (AMC) selection is essential for estimating accurate results in any area. The field-based hydrological measurements and soil sample collections provide data generation and mapping inventory. The systematic sampling according to slope and land use/land cover variation at micro-level assessment provides the topmost accuracy, results, and conservation plan for implementation. A watershed or drainage basin is a vital geomorphic unit for assessment, planning, and management. Sub-basin-wise evaluation and prioritization will ensure the sustainable management and development of land facets. This study model can be applied to evaluate the hilly terrain land systems of the Antecedent Moisture Condition (AMC) areas level.

References

Abdulwahd AK, Liejy MCh, Sulaiman MA, Al-Ansari N (2020) Water runoff estimation using geographical information system (GIS) for Alrakhmah Basin Valley Northeast of Iraq. Engineering 12:315–324. https://doi.org/10.4236/eng.2020.126025

Abraham M, Mathew RA (2018) Assessment of surface runoff for Tank Watershed in Tamil Nadu using hydrologic modeling. Hindawi, Int J Geophys 2018, Article ID 2498648:10p

Anjana SR, Jinu A (2019) Estimation of surface runoff using curve number method—a geospatial approach. Int Res J Eng Technol 6(9):1152–1157. e-ISSN: 2395-0056.

Askar MKh (2013) Rainfall-runoff model using the SCS-CN method and geographic information systems: a case study of Gomal River watershed. WIT Trans Ecol Environ 178:159–170. https://doi.org/10.2495/WS130141

Bhattacharyya R et al (2015) Soil degradation in India: challenges and potential solutions. Sustainability 7:3528–3570

Chatterjee C, Jha R, Lohani AK, Kumar R, Singh R (2001) Runoff curve number estimation for a basin using remote sensing and GIS. Asian Pac Rem Sen GIS J 14:1–7

Chow VT, Maidment DR, Mays LW (1988) Applied hydrology. McGraw Hill, New York

Czajkowski K, Lawrence PL (2013) GIS and remote sensing applications for watershed planning in the Maumee River Basin, Ohio. In: Geospatial tools for urban water resources, geotechnologies and the environment, p. 7. https://doi.org/10.1007/978-94-007-4734-0_8

Dahe PD, Deshmukh BB (2018) Estimation of annual runoff in indrāvati sub Basin of Godavari River using statistical approach. Int J Innov Res Adv Eng V:167–175. https://doi.org/10.26562/IJIRAE.2018.myae10079

Dhruvnarayan VV (1993) Soil and water conservation research in India. ICAR Krishi Anusandhan Bhavan, Pusa, New Delhi.

Gupta PK, Panigrahy S, Parihar JS (2011) Impact of climate change on runoff of the major river basins of India Using Global Circulation Model (HadCM3) projected data. J Indian Soc Remote Sensing 39(3):337–344.

Gupta PK, Chauhan S, Oza MP (2016) Modelling surface run-off and trends analysis over India. J Earth Syst Sci 125(6):1089–1102. https://doi.org/10.1007/s12040-016-0720-z

ICAR (2010) Degraded and wastelands of India status and spatial distribution. Published by Dr T P Trivedi, Project Director, Directorate of Information and Publications of Agriculture, Indian Council of Agricultural Research, Krishi Anusandhan Bhavan I, Pusa, and New Delhi, vol. 1, p. 167

Kumar N, Yamac SS, Velmurugan A (2015) Application of remote sensing and GIS in natural resource management. J Andaman Sci Assoc 20(1):1–6.

Kumar N, Tischbeinb B, Kuschec J, Lauxd P, Bege MK, Bogard JJ (2017) If Impact of climate change on water resources of upper Kharun catchment in Chhattisgarh India. J Hydrol Reg Stud 13:189–207. https://doi.org/10.1016/j.ejrh.2017.07.008

Matej VA, Vojtekovάb J (2019) Land use change and its impact on surface runoff from small basins: a case of Radiša basin. Folia Geographica 60(2):104–125

Michal F, Gabriela M (1997) Remote sensing, geographical information systems, snowmelt runoff models - an integrated approach. Remote sensing and GIS for design and operation of water resources system, vol. 242. IASH Publication, pp 73–82

Mohamed ES, Abdellatif MA, Abd-Elmabod SK, Khalil MMN (2020) Estimation of surface runoff using NRCS curve number in some areas in Northwest Coast, Egypt. E3S Web Conf 167:02002. https://doi.org/10.1051/e3sconf/202016702002. ICESD 2020

Nagarajun HJ, Kumar G, Surendra HJ (2016) Estimation of Runoff using SCS-CN and GIS method of ungauged watershed; a case study of Kharadya mill watershed, India. Int J Adv Eng Res Sci 3(5):36–42. https://doi.org/10.1155/2018/2498648

Narayana D (1993) Soil and water conservation research in India. Indian Council of Agricultural Research, Krishi Anusandhan Bhavan, Pusa, New Delhi, pp:146–15

Pawar DN, Sunil G (2013) Estimation of surface runoff of hilly terrain watershed using rainfall runoff modeling-a geoinformatics approach. Published in proceeding of Climate change and Human Response Aditya Prakashan Latur. pp. 1–4, ISBN 1-4978-93-83109-01-2

Putty MRY, Prasad R (2000) Understanding runoff processes using a watershed model - a case study in the Western Ghats in South India. J Hydrol 228:215–227. PII: S0022-1694(00)00141-4

Pandey AC, Stuti (2017) Geospatial technique for runoff estimation based on SCS-CN method in upper south koel river basin of Jharkhand (India). Int J Hydro 1(7):213–220. https://doi.org/10.15406/ijh.2017.01.00037

Pietroniro, A, Prowse TD (2002) Applications of remote sensing in hydrology. Hydrol Process 16:1537–1541 (2002) Published online in Wiley Inter Science (www.interscience.wiley.com). https://doi.org/10.1002/hyp.1018.

Rejani R, Rao KV, Osman M, Chary GR, Pushpanjali, Sammi Reddy K, Srinivasa Rao Ch (2015) Spatial and temporal estimation of runoff in a semi-arid microwatershed of Southern India. Environ Monit Assess 187:540. https://doi.org/10.1007/s10661-015-4772-3

Satheeshkumar S, Venkateswaran S, Kannan R (2017) Rainfall–runoff estimation using SCS–CN and GIS approach in the Pappiredipatti watershed of the Vaniyar sub basin South India. Model Earth Syst Environ 3:24

Savinai V (2016) Impact of temporal changes of land use on surface run-off -case study on Musi basin using Soil and Water Assessment Tool (SWAT). Report Submitted to JNTU, Hyderabad, International Crops Research Institute for the Semi-Arid Tropics Patancheru, Telangana, India, pp 2–80.

Schultz GA (1996) Remote sensing applications to hydrology: runoff. Hydrol Sci J 41(4):453–475. https://doi.org/10.1080/02626669609491520

SCS (1985) Chapter 10: Hydrology. In: National Engineering Handbook, Supplement A, Section 4. Soil Conservation Service, USDA, Washington DC.

Sharma A, Kanga S (2020) Surface runoff estimation of Sind River basin using SCS-CN method and GIS technology. https://doi.org/10.21203/rs.3.rs-41218/v1

Shivakoti BR, Fujii S, Tanaka S, Ihara H, Moriya M (2011) Remote sensing and GIS application for river runoff and water quality modeling in a hilly forested watershed of Japan. J Hydroinfor:198–216. IWA Publishing

Singh M, Shinde V, Pradhan SK, Chalodiya AL, Shukla KN (2014) Hydrologic modelling of Mahanadi river basin in india using rainfall-runoff model. Nat Environ Poll Technol Int Q Sci J 13(2):385–392. ISSN: 0972-6268

Singh P, Mishra SK (2017) Determination of curve number and estimation of runoff using Indian experimental rainfall and runoff data. J Spat Hydrol 13(1), Article 2

Sishah S (2021) Rainfall runoff estimation using GIS and SCS-CN method for awash river basin Ethiopia. Int J Hydro 5(1):33–37. https://doi.org/10.15406/ijh.2021.05.00263

Sun WC, Ishidaira H, Bastola S (2010) Towards improving river discharge estimation in ungauged basins: calibration of rainfall-runoff models based on satellite observations of river flow width at basin outlet. Hydrol Earth Syst Sci 14:2011–2022. https://doi.org/10.5194/hess-14-2011-2010

Tailor D, Shrimali NJ (2016) Surface runoff estimation by SCS curve number method using GIS for Rupen-Khan watershed, Mehsana District, Gujarat. J Indian Water Resour Soc 36(4)

Tripathi MP, Panda RK, Pradhan S, Sudhakar S (2002) Runoff modelling of a small watershed using satellite data and GIS. Photon J Indian Soc Remote Sens 30(1 & 2):39–52. https://doi.org/10.1007/BF02989975

Vojtek M, Vojtekova J (2016) GIS-based approach to estimate surface runoff in small catchments: a case study. Quaest Geogr 35(3):2016. https://doi.org/10.1515/quageo-2016-0030ISSN0137-477X

Wang X, Xie H (2018) A review on applications of remote sensing and GIS in water resources and flood risk management. Water 10:608. https://doi.org/10.3390/w10050608

Weng Q (2001) Modeling urban growth effects on surface runoff with the integration of remote sensing and GIS. Environ Manag 28(6):737–748. Springer-Verlag, New York Inc

Chapter 10
Valuing Benefits of Urban Green Spaces for Mitigation of Climate Change Impacts and Promoting Urban Resilience

Dhanya V. B., Sourav Dutta, Hemant Bherwani, Atya Kapley, Rajesh Biniwale, and Rakesh Kumar

Abstract Globally, increased urbanization and rapid population growth are exerting a burden on environmental resources, leading to climate change. Urbanization and population growth are however key components of society and it is important that resilience is built within the system to promote sustainable livelihood. Urban Green Spaces (UGS) along with public open spaces and common services, play a vital role for the metropolis to build a healthy surrounding. The diversity of roles (also known as Ecosystem services (ES)) of UGS, which vary from carbon sequestration to microclimate regulation, demonstrates their importance in maintaining environmental balance. The economic assessment of such services sheds light on the significance of UGS in climate regulation and allied services. The current study deals with the assessment of UGS available in Nagpur city and the changes in its land use/land cover (LULC) are studied from 1990 to 2020. A loss of 20.63 crores INR of ES has been assessed as a result of the research area's deteriorating urban green spaces during a 30-year period, which has harmed the social, economic, and, most crucially, the environmental balance. Using the results of the paper, the stakeholders and policymakers can make informed decisions for improving urban planning processes for long-term sustainability.

Keywords Ecosystem services · Urban-green spaces · Remote sensing · Climate regulation · Valuation

---

D. V. B. · S. Dutta · H. Bherwani (✉) · A. Kapley · R. Biniwale
CSIR-National Environmental Engineering Research Institute (NEERI), Nagpur, Maharashtra 440020, India
e-mail: h.bherwani@neeri.res.in

H. Bherwani · A. Kapley · R. Biniwale
Academy of Scientific and Innovative Research (AcSIR), Ghaziabad, Uttar Pradesh 201002, India

R. Kumar
Head Quarter, Council of Scientific and Industrial Research (CSIR), New Delhi 110001, India

A. Dubey et al. (eds.), *Promoting Sustainability Through Water Management and Climate Change Adaptation*, Advances in Geographical and Environmental Sciences,
https://doi.org/10.1007/978-981-99-5479-7_10

10.1 Introduction

Sustainable metropolises are erected on the foundation of healthy environs, which dominate and affect social welfare and most economic activities (Mader et al. 2010). Urban Green Spaces (UGS) is a vital share of these environs since it forms a metro's public open spaces and common facilities, and can help to promote a healthy atmosphere for all residents of an urban area and functions as a vital component in urban microclimate regulations (Cilliers et al. 2013; Bherwani et al. 2020a). It also serves as an ecological service supplier in metropolitan run-off regulation (Silvennoinen et al. 2017). Furthermore, these green areas have a profound influence on the physical and mental health of metropolitan residents (Bherwani et al. 2021). These are perceived as a refreshing break in the middle of bustling city life, and the Ecosystem Services (ES) they offer influence all sections of life, covering societal, financial, and environmental variables. Street trees, lawns/parks, urban forests, and cultivated land are examples of urban green areas (Bolund and Hunhammar 1999). Across many countries, the ES concept of various natural resource functionalities is progressively being used to demonstrate how biodiversity and ecosystems are involved in social well-being and must be emphasized in sustainability-based development in urban setup to establish a sustainable livelihood (Bherwani et al. 2020b). Alterations in the ecology have enhanced social welfare, trade, and industry, but they have also culminated in irreparable environmental damage (Aevermann and Schmude 2015). Intensified urbanization and rapid population growth are creating stress on these green resources, resulting in a considerable loss in urban green spaces today. Hence, the ecological benefits that these green spaces may supply are shrinking. The necessity of promoting UGS, its significance as well as the economic worth of the services they provide, can be better understood through monetary evaluation of their services (Aevermann and Schmude 2015). The assessment of these ecological services in urban settings offers information to stakeholders and decision-makers, allowing for better urban planning. Additionally, this information can be utilized to improve our understanding of how to preserve urban green spaces (Elmqvist et al. 2015). Hence in this research study, the UGS available in Nagpur city (Maharashtra) and the changes in its areal extent are studied for over 30 years using advanced tools like Remote Sensing (RS) and the monetary value of ecosystem services they provide is quantified. The location of study area is provided below (Fig. 10.1).

10.1.1 Objectives of the Study

The major objectives of the current research study are illustrated in the following (Fig. 10.2):

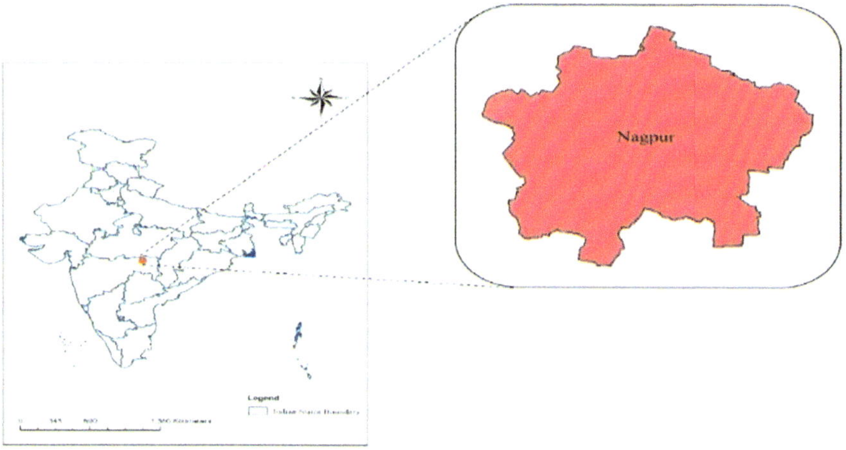

Fig. 10.1 Location map of study area (Nagpur)

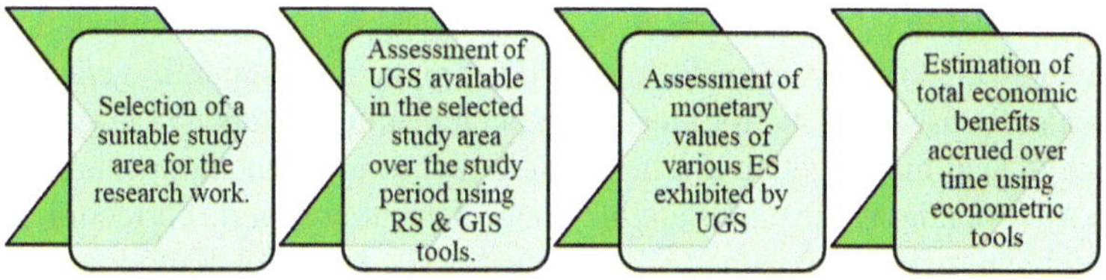

Fig. 10.2 Major objectives of the study

10.1.2 Methodology and Data Collection

10.1.3 Assessment of Urban Green Spaces of Nagpur City

The purpose of this research is to estimate the available areal extent of green spaces in Nagpur city and the variation in their area from 1990 to 2020. The approaches adopted for the research study are described below in more detail.

10.1.4 Land-Use and Land-Cover Maps (LULC Maps)

LULC maps (land-use and land-cover maps) are essential for resolving environmental challenges, project execution, and natural resource regulation. It also narrates human's actions on the ground and the different purposes to which it is put. Contrasting LULC maps from different years can help assess the total size of a geographical area used for development initiatives. The satellite images of Landsat

5 TM and Landsat 8 OLI for the years 1990 and 2020 were retrieved from the public domain US Geological Survey Earth Explorer Portal for the creation of LULC maps of Nagpur city. The data for the procedure were collected two months before and after the monsoon, to ensure that cloud cover will be minimal and the image obtained is clear for optimal accuracy. QGIS was used to perform Image Atmospheric Correction and Radiometric Correction & for the analysis and results, ArcGIS 10.6 was used.

10.1.5 Normalized Difference Vegetation Index Maps (NDVI Maps)

The greenness of vegetation, estimation of vegetation density, and discovering fluctuations in plant health can all be possible measures by the NDVI scale. It's a simple graphic representation for interpreting remote sensing data, typically from a space platform, to determine whether the viewing target contains live greenery. The Landsat 5 TM and Landsat 8 OLI data are predominantly valued in NDVI calculations, and they are collected for the construction of NDVI maps for Nagpur. In general, NDVI is a frequently used statistic for measuring plant health. High NDVI levels are associated with healthier vegetation. There is minimal or no vegetation when the NDVI is low. After obtaining the basic satellite image data, atmospheric correction and pre-processing are performed using QGIS software. The NDVI of Nagpur is calculated using the ArcGIS 10.6 platform. The LULC and NDVI maps for the year 1990 and 2020 are shown in figures below (Figs. 10.3 and 10.4).

10.1.5.1 Monetary Valuation of Ecosystem Services for Urban Green Spaces

As previously stated, the UGS provides enormous ecological services to the city residents. Provisioning, regulating, supporting, and cultural services are the four types of services available. A variety of research articles were analyzed to collate the values of various service categories. A wide variety of ecological services provided by UGS are studied, and financial benefits for these functions are derived from various literature. Those data are subsequently transferred to the research area using the "Value Transfer Method" (VTM). The VTM was created to determine the monetary value of the services, and it is based on the concept of data transfer from previously completed research to new sites or contexts.

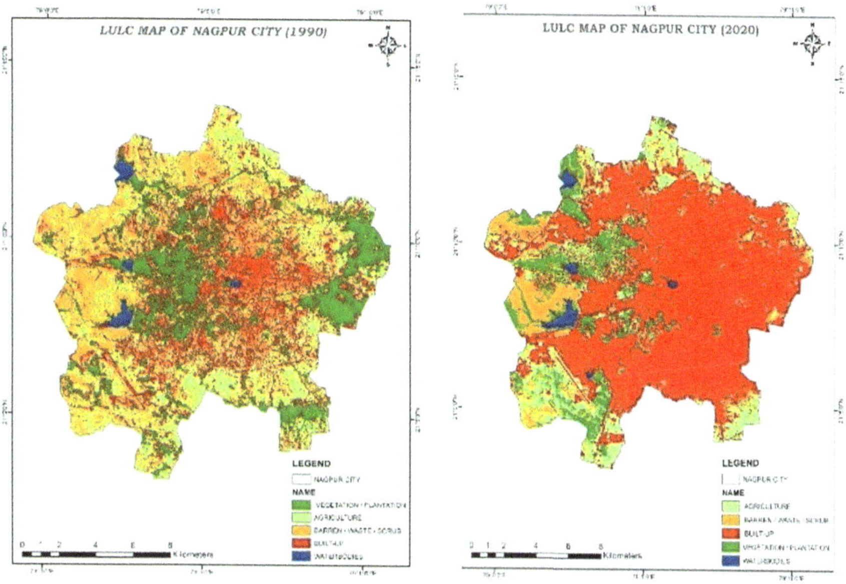

Fig. 10.3 LULC maps of Nagpur city for the year 1990 & 2020

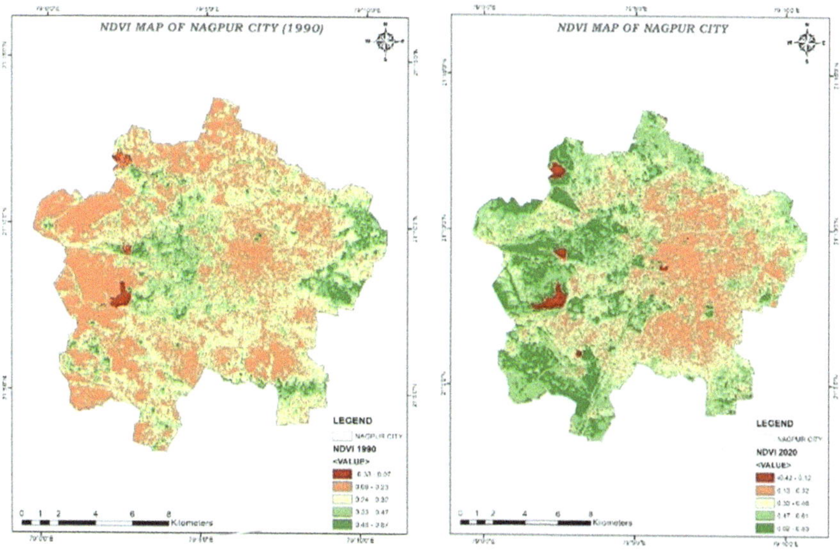

Fig. 10.4 NDVI maps of Nagpur city for the year 1990 & 2020

Table 10.1 Details of LULC maps for the years 1990 and 2020

Classification	1990		2020	
	Area (sq. km)	Area (%)	Area (sq. km)	Area (%)
Vegetation/plantation	40.11	18.19	21.07	9.56
Waterbodies	2.46	1.12	2.77	1.26
Built-up	61.66	27.97	137.42	62.32
Agriculture	57.07	25.88	32.75	14.85
Barren/waste/scrub	59.18	26.84	26.48	12.01
Total area	220.49	100.00	220.49	100.00

10.2 Results & Discussion

The various results obtained from the above-mentioned objectives are briefly described in the following section.

10.2.1 Assessment of Urban Green Spaces for Nagpur City

10.2.1.1 Results from LULC Maps of Nagpur City

The following Table 10.1 provides the analysis results of LULC maps prepared for the years 1990 and 2020.

The above table demonstrates that between 1990 and 2020, there is a significant loss in UGS (Vegetation/Plantation) in the studied area. It was 40.11 sq. km in 1990 and 21.07 sq. km in 2020, indicating a decline of 19.04 sq. km of greenery during 3 decades that must be properly considered. On the other side, the built-up area expanded from 61.66 sq. km in 1990 to 137.42 sq. km in 2020, which could be a factor in the shrinking of these green spaces.

10.2.1.2 Results from NDVI Maps of Nagpur City

The details of analysis results obtained from the NDVI maps of Nagpur are given in Table 10.2 below.

According to the evaluation from the above data, the research region's vegetative area has decreased significantly from 1990 to 2020. It was 97.18 sq. km in 1990 and 53.82 sq. km in 2020, indicating a loss of 43.36 sq. km of the vegetative area over 30 years.

Table 10.2 Details of NDVI maps for the years 1990 and 2020

Classification	1990		2020	
	Area (sq. km)	Area (%)	Area (sq. km)	Area (%)
Non-vegetative area	123.31	55.93	166.67	75.59
Vegetative area	97.18	44.07	53.82	24.41
Total area	**220.49**	**100.00**	**220.49**	**100.00**

10.2.1.3 Monetary Valuation of Ecosystem Services for Urban Green Spaces

The above-mentioned methodology is used to acquire and assess the reference monetary values of services provided by the UGS of the study area from various literature studied, including the research works from Spain, the United States, China, Canada, England, and Germany. Various services like climate regulation, recreation, carbon sequestration, air quality regulation, runoff reduction, water flow regulation/groundwater recharge, temperature control/energy conservation, O_2 generation, microclimate control, and other services are considered and the reference values for the same are procured from the reviewed literature (Elmqvist et al. 2015; TEEB Valuation database 2010; Xie et al. 2019; Dennis and James 2016). The values thus selected are inflated and suitable Market Exchange Rates (MER) are used in order to estimate the ecosystem service values of Nagpur city for the year 2020. Since the services rendered by the various ecosystems are of enormous significance and are always compared with market values it is appropriate to take MER for monetary evaluation. The average of the values inflated considering the above-mentioned services are determined and the final monetary value of ecosystem services is assessed for the study area for the year 2020 and it is found to be 1,08,327 INR/ha/year. As stated above the total reduction in the UGS of Nagpur city is 19.04 sq. km (1904 ha) and the total monetary value of the services they provide is determined using the above derived monetary value of ecosystem services and the yielded estimates are provided in the following Table 10.3.

As shown above, the decline of green spaces has been estimated to be 1904 ha, with the total economic value of ecological services estimated to be 20.63 crores INR in 2020. This must be regarded as a bigger loss caused by the decline of urban green spaces because it will reflect its influence on community, financial, and ecologic facets of society.

Table 10.3 Monetary value of urban green spaces of Nagpur city

Total reduction in urban green spaces from 1990–2020 (ha)	The average value of ecosystem services (INR/ha/year)	Total value (Crore INR)
1904	1,08,327	20.63

10.3 Conclusion

The idea of sustainability is extremely significant in the present scenario of tremendous population growth and unchecked resource usage. Variations in the climatological conditions are enormous as a result of unregulated resource utilization. Resource-intensive development for human well-being has evolved into a curse for life's existence. For the sake of future generation's well-being, the notion of sustainable livelihood should be a current top priority, so that the balance between ecology and economy and its positivity will reflect in the future as well. The soul of megacities is urban green space, which provides a welcome relief from the busyness of daily life. Furthermore, the value of the services they deliver are immense. This analysis of Nagpur's urban green areas demonstrates that the area of these spaces is shrinking year after year. The boom in urbanization and population increase, resulted in an increase in the built-up area of the city and a reduction in available green spaces. The ecological services offered by Nagpur's degraded green space are worth a total of approximately 20.63 crores INR. These lost services had the potential to increase urban resilience in the face of severe climate change and to improve sustainable livelihoods. Strict rules and procedures should be implemented to ensure that these green places are properly preserved and that the full capacity of their contributions to society is achieved.

References

Aevermann T, Schmude J (2015) Quantification and monetary valuation of urban ecosystem services in Munich Germany. Zeitschrift Für Wirtschaftsgeographie 59(3):188–200

Bherwani H, Anju Singh A, Kumar R (2020a) Assessment methods of urban microclimate and its parameters: a critical review to take the research from lab to land. Urban Clim 34:100690. https://doi.org/10.1016/j.uclim.2020.100690

Bherwani H, Indorkar T, Sangamnere R, Gupta A et al (2021) Investigation of adoption and cognizance of urban green spaces in India: post COVID-19 Scenarios. Curr Res Environ Sustain 3:100088. https://doi.org/10.1016/j.crsust.2021.100088

Bherwani H, Nair M, Kapley A, Kumar R (2020b) Valuation of ecosystem services and environmental damages: An imperative tool for decision making and sustainability. *Eur J Sustain Dev Res* 4(4):em0133. https://doi.org/10.29333/ejosdr/8321

Bolund P, Hunhammar S (1999) Ecosystem services in urban areas. *Ecol Econ* 29(2):293–301. https://doi.org/10.1016/S0921-8009(99)00013-0. Available at: http://uknea.unepwcmc.org/EcosystemAssessmentConcepts/EcosystemServices/tabid/103/Default.aspx

Cilliers S, Cilliers J, Lubbe R, Siebert S (2013) Ecosystem services of urban green spaces in African countries—perspectives and challenges. Urban Ecosyst 16:681–702. https://doi.org/10.1007/s11252-012-0254-3

Dennis M, James P (2016) Considerations in the valuation of urban green space: accounting for user participation. *Ecosyst Serv* 21(Part A):120–129. https://doi.org/10.1016/j.ecoser.2016.08.003

Elmqvist T, Setala H, Handel SN, Ploeg SD et al (2015) Benefits of restoring ecosystem services in urban areas. Curr Opin Environ Sustainab 14:101–108. https://doi.org/10.1016/j.cosust.2015.05.001

Mader A, Patrickson S, Calcaterra E, Smit J et al (2010) Teeb manual for cities: ecosystem services in urban management. *Econ Ecosyst Biodiv.*
Silvennoinen S, Taka M, Yli-Pelkonen V, Koivusalo H, Ollikainen M, Setälä H (2017) Monetary value of urban green space as an ecosystem service provider: a case study of urban runoff management in Finland. Ecosyst Serv 28:17–27. https://doi.org/10.1016/j.ecoser.2017.09.013
TEEB Valuation database (2010). https://www.cbd.int/financial/gmr/teeb-database.xls
Xie Q, Yue Y, Sun Q, Chen S et al (2019) Assessment of ecosystem service values of urban parks in improving air quality: a case study of Wuhan China. Sustainability 11(22):6519. https://doi.org/10.3390/su11226519

Chapter 11
Reinforcement of Drinking Water in Fluoride Affected Areas of Nalgonda District Through Improvised Rainwater Harvesting System

M. V. S. S. Giridhar and Shyama Mohan

Abstract Rainwater harvesting is a simple technique that allows rainwater to percolate into the soil. Rainwater harvesting must be considered in areas with high fluoride concentrations in the groundwater in order to make the groundwater drinkable and for other domestic reasons. The primary objective of the study is to demonstrate in practice the efficacy of modified Rainwater Harvesting (RWH) systems in lowering the fluoride concentration in a community where fluoride levels are more than 6 ppm at a depth of 180 ft below ground level. An enhanced Rainwater Harvesting System was developed at the ZPHS School in Nalgonda's Gottiparthi hamlet, Telangana, India, utilizing a specialized four-layer filter media. A pre- and post-monsoon fluoride assessment was also carried out for four years since the construction of the RWH structure. According to the study, RWH structures were shown to be effective, significantly decreasing fluoride levels in groundwater over the period of four years. Prior to the monsoon, the maximum fluoride concentration was 7.45 ppm in 2017 at the School Premises, reducing to 1.23 ppm during the post-monsoon in 2020. Fluoride concentrations in freshly dug RWH borewells of depth 100 ft and 200 ft were 5.16 ppm and 5.30 ppm, respectively, before construction in 2017 and had reduced to 0.70 and 1.04 ppm in the post-monsoon season of 2020.

Keywords Nalgonda · Fluoride · Rainwater harvesting · Water quality · Fluorosis

M. V. S. S. Giridhar (✉) · S. Mohan
Centre for Water Resources, Institute of Science and Technology, Jawaharlal Nehru Technological University Hyderabad, Kukatpally, Hyderabad, Telangana 500085, India
e-mail: mvssgiridhar@gmail.com

A. Dubey et al. (eds.), *Promoting Sustainability Through Water Management and Climate Change Adaptation*, Advances in Geographical and Environmental Sciences,
https://doi.org/10.1007/978-981-99-5479-7_11

11.1 Introduction

Fluoride occurs naturally in the environment and people of the region consume it in diminutive amounts. The diminutive amounts can occur through dietary intake, respiration, and fluoride supplements. Fluoride is highly reactive element and ranks 13th in terrestrial abundance and represents 0.06–0.09% of weight of the Earth's crust. Fluorine is the natural component present in lithosphere, atmosphere, hydrology, and biosphere that is present. Significant amount of fluoride can be found in volcanic rocks. Rocks high in fluoride are the primary source of fluoride in groundwater. In different parts of the world, a high concentration of fluoride was found in ground water and surface water (Sakthi et al. 2014). Rocks like Granite in certain areas form a major reason for fluoride contamination in groundwater. 65 million citizens in India are at risk for dental fluorosis and skeletal fluorosis in various states in India are suffering from endemic fluorosis like Telangana, Karnataka, Andhra Pradesh, Rajasthan, Tamil Nadu, Haryana, Gujarat, Bihar, Kerala, and Punjab. The granites in the districts like Nalgonda contain much higher fluoride than the world average fluoride concentration of 810 mg/kg. Hence the intake of fluoride via drinking water is the maximum in these regions. The permissible limit of fluoride in drinking water has been laid down as 1.5 mg/L by WHO and Indian Water Standard. Small amount of fluoride is required for the growth and strength of bones and teeth. But continuous intake of fluoride above the permissible limit causes severe skeletal and dental deformities. The Dental Flourosis prevalence in Nalgonda district is about 71.5% between the age group of 12–15 years and 30.6% in adults, whereas skeletal fluorosis prevalence is 24.9% in all age groups. Almost millions of people are affected by fluorosis. In recent years, rapid population growth, industrialization, intense agricultural activity, low rainfall, diminishing surface water resources, and climate change have caused severe water stress, especially in Andhra Pradesh, Telangana, and other rural parts of the country (Adimalla et al. 2019).

Hence, with increase in water stress, it has become a necessity to treat drinking water contaminated with high concentration of natural minerals like fluoride. There are several defluorination techniques for reduction of fluoride in drinking water. Nalgonda technique, adsorption technique, ion exchange technique, reverse osmosis, and electro dialysis are some of the techniques famously employed for defluorination. These techniques usually require trained operator, daily operation, and are costly. These types of techniques also tend to produce sludge which poses serious environmental health problems. A sustainable approach for treating the groundwater directly is rainwater harvesting system. Rainwater harvesting will also ease the demand for conventional water supplies, lessen non-point source contamination loads, monitor water logging issues, prevent floods, aid in reducing the effects of climate change, and help manage stormwater, and other difficulties (Zhu et al. 2004). Rainwater harvesting seems to be a perfect replacement for surface and groundwater as latter is concerned with the rising cost as well as ecological problems. Thus, rainwater harvesting is a cost-effective and relatively lesser complex way of managing our limited resources ensuring sustained long-term supply of water to the community (Giridhar et al. 2011; Fewkes 1999; Gould and Nissen-Petersen 1999; Leggett et al. 2001; Heggen 2000). In India, evidence has been found of simple stone-rubble structures for impounding water that date back to the third millennium BC (Agarwal & Narain, 1997). Numerous research on impact of rainwater harvesting system on groundwater quantity and quality have been carried out. Brindha Karthikeyan et al, (2016) have studied fluoride in weathered rock aquifers of Southern India and observed impact of managed Aquifer Recharge for mitigation. Srinivas et al, (2008) have studied the impact of groundwater recharge through rainwater harvesting and variation in water quality. The study depicts that influence of check dams on water quality in bore wells near it is more and its influence gradually decreases with distance. Giridhar M.V.S.S., and K. Divya Chowdary (2015) showed the effectiveness of Rainwater recharge into ground water aquifers through recharge Shafts in balancing the fluoride concentrations of ground water and the roof top runoff that is free of fluorides contributes effectively in diluting fluoride concentrations.

The objective of present study is to construct the improvised rainwater harvesting structures in school premises for artificial recharge of groundwater and to analyze groundwater samples during pre- and post-monsoon period in the fluoride-affected groundwater.

11.2 Study Area

Gottiparthi is a hamlet located in Thungathurthi Block of Nalgonda district in Telangana. Situated in rural part of Nalgonda district of Telangana, it is one among the 20 villages of Thungathurthi Block of Nalgonda district. Gottiparthi has a total population of 4109 peoples. The total geographical area of Gottiparthi village is 14 km^2 and it is the 5th biggest village by area in the sub-district. It is situated between 7°35′1.9″N Longitude 17°29′5.46″E latitude and 79°35′37.3″N longitude 17°29′41.8″E Latitude. The location of Gottiparthi village is shown in Fig. 11.1. The rainwater structures are constructed in the Zilla Parishadh High School (ZPHS) situated in the Gottiparthi Village. The ZPHS School covers an area of 3.29 acres. The location of borewells and the school is shown in the Fig. 11.1.

11.3 Methodology

The main objective of the study is to construct the improvised rainwater harvesting system and prove its efficiency from the groundwater quality analysis. The methodology is explained in further sections.

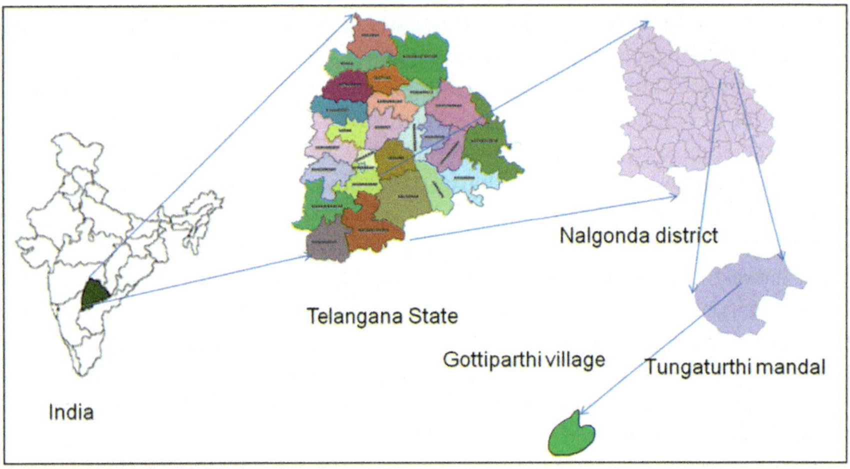

Fig. 11.1 Location map of the Gottiparthi village in Nalgonda district, Telangana

11.3.1 Implementation of Improvised Rainwater Harvesting Structures

The site for the construction of rainwater harvesting structures at the ZPHS School was first identified. The lowest site suitable for the construction of rainwater harvesting structures was identified from elevations obtained from the GPS device. The dimensions of the RWH structure are designed based on surface runoff computed for the selected study area. The computation of surface runoff was analyzed from the average rainfall and area of the catchment. The latitude and longitudes of the site were recorded and noted as shown in Table 11.1 following the identification of the site, the rain water harvesting structures were constructed with two recharge shafts in the ZPHS School of Gottiparthi village. This type of recharge structure is surface RWH, it captures the surface runoff from the catchment area and injects it below the ground surface. The RWH system implemented in this study is improvised version of traditional RWH pit.

Table 11.1 Location and details of the structures

Sr. no.	Bore wells	Location	Latitude	Longitude
1.	Newly drilled borewell 1	In the School Premises	17°29′ 21.8″	79°35′ 19.9″
2.	Newly drilled borewell 2	In the School Premises	17°29′ 21.8″	79°35′ 20.2″
3.	Borewell 3	In the School Premises	17°29′ 22.7″	79°35′ 17.0″
4.	Upstream-borewell 4	Behind the School	17°29′ 19.7″	79°35′ 17.0″
5.	Downstream-borewell 5	Outside the School Premises	17°29′ 23.5″	79°35′ 31.4″

The main component of the improvised RWH system is two recharge shafts along with the filter media. Two recharge shafts were drilled at different depths, i.e., 100 ft and 200 ft with supporting structures for these shafts. The casing had been fixed upto the hard strata and one-inch holes were drilled around the casing pipe for a depth of 3.5 ft. After which the four-layer mechanism of filter media was laid as shown in Fig. 11.2. In the four-layer mechanism the bottom-most layer is the 40 mm gravel which was filled upto depth of 1 ft followed by the second layer filter media consisting of 20 mm gravel which was layered upto the depth of 1 ft. Then the third layer consisting of graded coarse sand was filled upto the depth of 1 ft. The fourth and foremost layer is the unique layer of geo-membrane textile having perforations of 125 microns laid above the sand filter media. This geo textile prevents the dust particles and floating material to enter into the aquifer, only allowing the passage of rainwater effectively. The remaining one-fourth of the recharge pit, the top-most portion is left vacant in order to store excess rainwater which flows from the upstream catchment (Giridhar 2017). After the construction of the structures, the assessments of the groundwater quality for the presence of fluoride were carried out for the pre-monsoon and post-monsoon period.

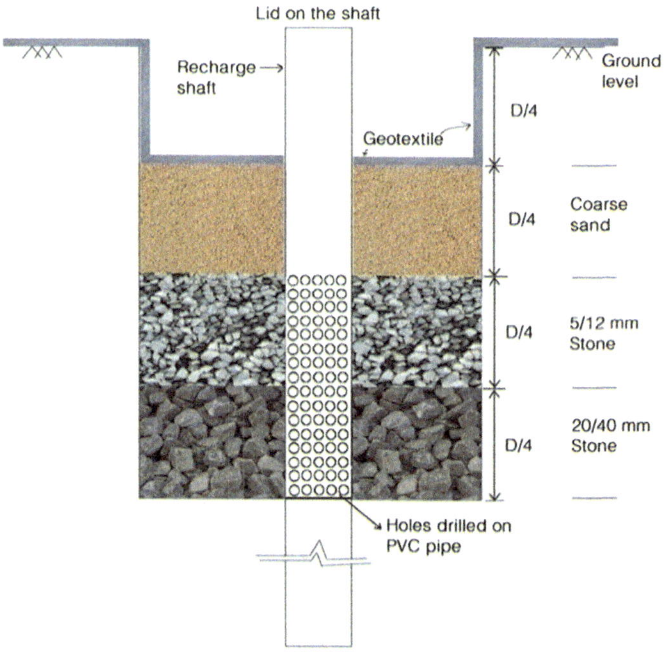

Fig. 11.2 Four-layer mechanism

11.3.2 Groundwater Quality Analysis

The groundwater samples were collected from five locations in the study area from the year 2017 to 2020. Samples from the borewells of two shafts (100 and 200 ft), from the upstream, downstream catchment, and within the school premises are further analyzed for the presence of fluoride. The samples were collected in sterilized plastic bottles. At each location, a borewell is selected from which the groundwater samples were collected for the entire period of study. Two bottles of samples from each borewell location point are collected and analyzed. A total of twenty samples of water bottles were obtained for every year (pre-monsoon and post-monsoon) which further resulted in 70 bottles for the period of four years for analysis of fluoride.

After the collection of samples, they are tested for fluorides. The protocol IS 3025/ Part 60 was followed for testing of 70 samples for fluoride. The results obtained for fluoride concentration will be then averaged at each location resulting in five readings for each season. The values were then digitized in MS Excel and graphs were plotted and the trend for fluoride concentration is analyzed.

11.4 Results and Discussions

After the construction of rainwater harvesting structure, the analysis of fluoride concentrations available in the groundwater is effectuated. The analysis was carried out for both pre-monsoon and the post-monsoon period from 2017 to 2020. The fluoride concentrations were obtained from four locations of the borewells. The samples had been collected from the two newly drilled borewells of rainwater structure, one borewell in the upstream present behind the school premises, another from outside the school premises present at the downstream and borewell within the school premises. The fluoride concentration for the pre-monsoon taken on 24th May 2017 from the samples showed a higher concentration than the desirable limit (1.5 mg/L). The highest fluoride concentration, 7.45 ppm was shown in the sample of borewell collected from school premises and the lowest in the newly drilled borewell 2 beside the school showing 5.30 ppm. The analysis for the post-monsoon period, i.e., on September 20, 2020 showed a favorable decrease in the concentration of fluorides. The highest concentration which earlier was 7.45 ppm before the monsoon period had decreased to 1.23 in post-monsoon period of the year 2020. The newly drilled borewells showed the concentration of fluoride 5.16 and 5.09 ppm in the beginning of construction of RWH which decreased to 0.70 ppm and 1.04 ppm respectively in the year 2020. The fluoride concentrations of the five samples are shown in Table 11.2 and its graphical representation is shown in Fig. 11.3.

Table 11.2 Fluoride concentration in PPM for pre and during monsoon of different bore wells

Sr. no.	Location of the borewell	Fluoride concentration in PPM						
		May 2017 (Pre monsoon)	September 2017 (Post monsoon)	April 2018 (Pre monsoon)	September 2018 (Post monsoon)	March 2019 (Pre monsoon)	October 2019 (Post monsoon)	September 2020 (Post monsoon)
1.	School premises	7.45	6.28	2.54	1.36	1.96	1.68	1.23
2.	Upstream	6.23	5.45	4.67	2.28	2.12	2.04	1.94
3.	Downstream	5.09	4.36	2.72	1.04	1.11	1.03	0.94
4.	Newly drilled borewell—1 at RWH	5.16	4.08	3.61	3.15	2.67	1.34	0.70
5.	Newly drilled borewell—2 at RWH	5.30	4.10	4.01	2.57	3.04	1.48	1.04

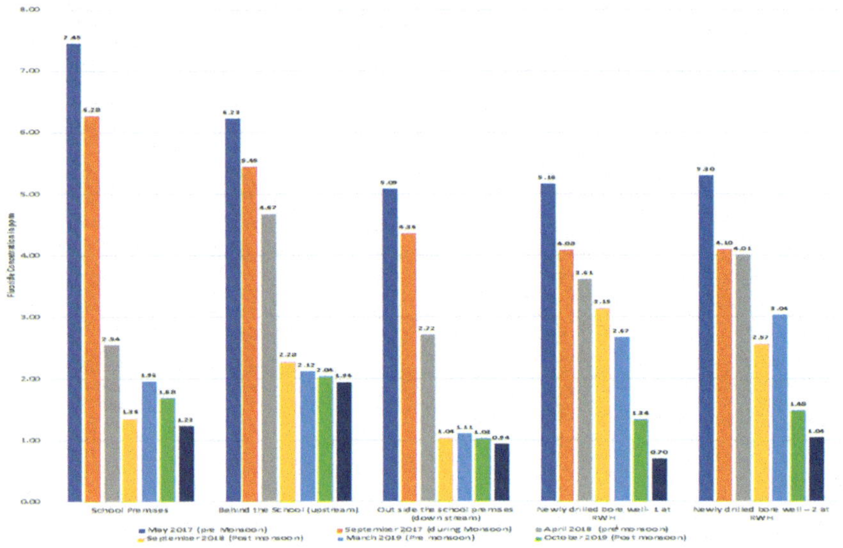

Fig. 11.3 Fluoride concentration in study area during pre & post-monsoon seasons for each location from the year 2017 to 2020

11.5 Conclusion

This research assessed the viability of rainwater harvesting with improvised structure and its influence in the fluoride-affected region of Nalgonda. The structure built in the school was found to be effective for the surrounding villages around the school. The groundwater quality near the downstream which is located at a distance of 100 m shows a tremendous decrease in fluoride concentration within a span of four years. From the graphical representations, the region around the RWH structure demonstrates reduced fluoride concentrations. This depicts that the influence of rainwater recharge on fluoride concentration is proportional to the distance from point of recharge. The pre-monsoon period in the year 2018 shows a drastic reduction in the concentration which is unusual for the quality of groundwater in Nalgonda region especially during the summer season. It is also seen at the recharge structure that the fluoride concentration is lesser at shallow depth of 100 ft than at 200 ft depth. This indicated the effectiveness of rainwater recharge into groundwater aquifers through recharge shafts in balancing the fluoride concentrations of groundwater. It was found that the amount of harvested rainwater near the premises of school has recharged the dry borewells that have been not functioning since long.

Misconception about rainwater harvesting is the main reason that people do not usually practice such RWH system. Hence, there is a need for awareness for the people to adopt the rainwater system for reduction of any water problems like groundwater quality or quantity.

References

Adimalla N, Venkatayogi S, Das SVG (2019) Assessment of fluoride contamination and distribution: A case study from a rural part of Andhra Pradesh, India. Appl Water Sci 9, 94. https://doi.org/10.1007/s13201-019-0968-y

Agarwal A, Narain S (1997) Dying wisdom: rise and fall of traditional water harvesting systems. Centre for Science and Environment, New Delhi, India

Brindha K, Jagadeshan G, Kalpana L, Iango L (2016) Fluoride in weathered rock aquifers of southern India: Managed Aquifer Recharge for mitigation. Environ Sci Pollut Res 23(9), 8302–8316. https://doi.org/10.1007/s11356-016-6069-7

Fewkes (1999) Modelling the performance of rainwater collection systems: towards a generalized approach. *Urban Water* 1:323–333

Giridhar MVSS (2017) Chapter 26: An environmentally sound approach for sustainable rainwater harvesting. In: Srinivasa Raju K, Vasan A (eds) Sustainable holistic water resources management in a changing climate. M/S Jain Brothers, New Delhi, pp 26.451–26.467

Giridhar MVSS, Divya Chowdary K (2015) Spatial distribution of ground water quality after rain water recharge in JNTUH campus. Int J Sci Res Eng Technol 4(Special issue 1). eISSN: 2319-1163 | pISSN: 2321-7308

Giridhar MVSS, Chandra Bose AS, Viswanadh GK (2011) Identification of suitable locations for rooftop rainwater harvesting structures. Int J Appl Sci Eng Res 1(1):101–106. ISSN: 2277-8442

Gould J, Nissen-Petersen (1999) Rainwater catchment systems for domestic supply: design, construction and implementation. Intermediate Technology Publications Ltd., London, UK

Heggen RJ (2000) Rainwater catchment and the challenges of sustainable development. Water Sci Technol 42(1,2): 141–145

Leggett DJ, Brown R, Brewer D, Stanfield G, Holliday E (2001) Rainwater and greywater use in buildings: best practice guidance. *CIRIA report C539*. Construction Information & Research Information Association, London, UK

Sakthi TA, Jeya KR, Rajakumar S, Ayyasamy PM (2014) "Incidence and effects of fluoride in Indian natural ecosystem: A review". Adv Appl Sci Res, 5(2):173–185, ISSN: 0976-8610

Srinivas P, Kumar P, Chowdary P, Kumar P (2008) Ground water recharge through rainwater harvesting and studying variation in water quality. Enviromed Roll Res 27(3):503–506

Zhu K, Zhang L, Hart W, Liu M, Chen H (2004) Quality issues in harvested rainwater in arid and semi-arid Loess Plateau of northern China. J Arid Environ 57(4):487–505

Chapter 12
Assessment of the Farmer Support Initiative Programmes in Telangana State—A Case of Mission Kakatiya and Rythu Bandu

T. Anuradha and R. Sudhakar Goud

Abstract After formation of Telangana state, the government has taken major initiative programmes such as; Mission Kakatiya, lift irrigation projects and Rythu Bandu and Rythu Bheema. In order to overcome the drought situation, the schemes are playing a key role across the state. More than 40,000 tanks have been renovated and developed in four phases, under the Mission Kakatiya Scheme. The state government, nearly Rs. 15,000 crore rupees are spending on the Rythu Bandu scheme, every year and very recently (January 2021) a total of Rs. 7,515 crore rupees were deposited in 61.49 lakh farmers' accounts. Based on the performance of the schemes, Central institutions like; NITI Ayog, NABARD and other international organizations are also praised. The primary data was collected through Questionnaire and interview method. In this regard, a total of 330 farmers were interviewed, in the state. Based on the primary data, the paper concluded that, Mission Kakatiya programme resulted in crop productivity, soil fertility and ground water levels have been increased and also created employment. Overwhelming percentage of the eligible farmers are benefitting under Agriculture Investment Support Scheme (Rythu Bandu). Based on the views of the farmers, the study suggests the need to integrate agriculture crop works to 100 days work programme and extend Rythu Bandu to land tenants (Kaulu Rythulu).

Keywords Mission Kakatiya · Lift irrigation projects · Rythu Bandu and MGNREGs

---

T. Anuradha (✉)
Department of Geography, Osmania University, Telangana State, Hyderabad 503007, India

R. S. Goud
Department of Geoinformatics, Telangana University, Telangana State, Nizamabad 503322, India

A. Dubey et al. (eds.), *Promoting Sustainability Through Water Management and Climate Change Adaptation*, Advances in Geographical and Environmental Sciences,
https://doi.org/10.1007/978-981-99-5479-7_12

12.1 Introduction

In our state more than 70% of the rural population depends on Agriculture and Allied activities. When farmer suffers from unexpected natural disasters like failure of monsoons (low rainfall-drought) and heavy rainfall, floods, etc., then agriculture sector gets affected by less yield and productivity. In the drought situation majority of the wells, ponds lakes, or reservoirs levels sinks and farmer livelihood gets affected. During this particular time, the farmer needs better support from the Government and for his existence and survival, to continue agriculture.

Perennial rivers like Godavari and Krishna are facilitating huge amount of water. Generally, the reservoirs are constructed for multiple purposes usage, i.e., irrigation and drinking water. Availability of water is not enough for the farmer; they need economical support to cope up with their life. Our state and central government have been implementing many farmers-based schemes, i.e., free power, subsidy-based fertilizers, agriculture tools, credit facility for farmers, crop insurance schemes, tractors to farmers, PM Kisan Mann Dhan Yojana, Pradhan Mantri Kisan Nidhi, Pradhan Mantri Krushi Sinchai Yojana, etc. Among these schemes, some of them got benefited.

Among all, one of the recent initiatives is Pradhan Mantri Krishi Sinchai Yojana (PMKSY) and PM Kisan Scheme. These two schemes are very important and helpful to the country farmers. Under PMKSY, every agriculture field gets water. It is also called "Har Khet Ko pani," which aims to improve water use efficiency by more crops per drop. Another scheme is PM Kisan Scheme, which is providing Rs. 6,000/- per year as a minimum income to the less than two hectares of landholding farmers. In a similar way Telangana state is also providing schemes to farmers but in this study, we consider only Mission Kakatiya which is renovation of tanks and Rhythu Bandu programmes to support farmers economically in the state.

12.2 Objectives of the Study

The main objective of this paper is to know the progress of work under Mission Kakatiya and to assess the implementation of farmer support initiative Rythu Bandhu programme.

12.3 Methodology

This paper is analyzed on primary and secondary data basis. The secondary data has been taken from Telangana Statistical Abstract and Socio Economic Outlook of Telangana state. The primary data was collected through the questionnaire, which was prepared. A total of 300 farmers were taken as samples (100 samples from each agro-climatic region of Telangana, i.e., Northern, Central and Southern agro-climatic regions) for the study in various districts of Telangana. The secondary data was collected and data tabulated and generated tables and charts.

12.4 Context of Study Area

Telangana state has 1, 12,077 sq. km area and having 3.5 crore people. Nearly 70% of people are engaged in the primary activity. The state receives a maximum of 80% rainfall during the southwest monsoon period and the remaining 20% received in the other seasons. It varies from district to district, which is received between 800 and 1200 mm.

Previously, the state received deficit rainfall. During this deficit rainfall period, the farmers of Telangana faced a lot of problems and many farmers committed suicide; because they were depending on the private financiers for the agriculture investments at high interest rates. Due to the loss of crop, they didn't repay the financiers. Hence the Telangana state government has introduced many farmers-supported schemes across the state. They are like; Mission Kakatiya, Haritha Haram Rythubandu and Rythubheema. In this context schemes like Mission Kakatiya and Rythubandu are analyzed below.

12.5 Results and Discussion

Status and work progress of Mission Kakatiya programme since the formation of Telangana: This programme was started in March 2015 which is the first programme to be taken by the Government of Telangana. The main objective of Mission Kakatiya is to enhance the agriculture-based income for small and marginal farmers, by accelerating the development of minor irrigation infrastructure, strengthening community-based irrigation management and adopting a wide-ranging programme for restoration of tanks. A total of 60,000 tanks were surveyed in the state. In Siddipet district, there are about 4,848 tanks, having the highest number of tanks followed by Medak, Bhadradri, Mahabubnagar, Nalgonda and Rangareddy districts, the lowest number of tanks are surveyed in Adilabad, Rajanna, Hanamkonda, Jogulamba, Medchal and Hyderabad in the given Table 12.1.

The Mission Kakatiya programme was implemented phase wise in the state. A total of 7,991 tank works were started, in first phase and almost 93% of tank works were finished. In second phase nearly 75% tanks; in third and fourth phase, 43 and 32% of the commenced tanks works have completed. It is observed 40% (18,738) tanks works have been completed, in the state (Table 12.2 and Fig. 12.1).

It was observed that the Gross Area Irrigated (GAI) increased from the 53.2 lakh hectares to 113.2 lakh hectares from 2014–2015, to 2019–2020, During the same period, there is good improvement in the Net Area Sown which has increased from

Table 12.1 District wise tanks and its Ayacut details in Telangana State

District	Survey completed	Ayacut (Acres)
Siddipet	4,848	146,112
Medak	3,716	115,231
Bhadradri	3,705	225,965
Mahbubnagar	3,143	211,825
Nalgonda	3,113	160,958
Rangareddy	3,007	64,837
Sangareddy	2,964	126,352
Nagarkurnool	2,837	98,657
Jayashankar	2,797	158,386
Kamareddy	2,468	118,725
Mahabubabad	2,191	133,967
Yadadri	1,904	109,897
Khammam	1,838	161,456
Wanaparthy	1,790	81,650
Karimnagar	1,761	74,840
Peddapalli	1,602	119,446
Jagtial	1,524	100,818
Suryapet	1,520	91,638
Nizamabad	1,483	142,847
Vikarabad	1,374	87,898
Warangal R	1,273	87,767
Jangaon	1,216	68,746
Mancherial	1,131	109,151
KB-Asifabad	1,115	95,975
Nirmal	1,054	108,607
Adilabad	870	111,057
Rajanna	841	53,217
Warangal U	841	49,752
Jogulamba	815	23,680
Medchal	561	22,538
Unmapped	12	10
Hyderabad	6	105
Total	59,320	3,262,112

Source Mission Kakatiya, Telangana State, 2019

Table 12.2 Details of phase wise covered tanks and its status in Telangana State

Phase wise	Total tanks	Total GO issued	Works commenced	Works completed	% of works completed to work commenced
Phase I	46,558	8,219	7,991	7,660	93.2
Phase II	46,558	9,286	8,975	6,981	75.2
Phase III	46,558	6,133	5,789	2,634	42.9
Phase IV	46,558	4,626	3,711	1,463	31.6
Grand Total		28,264	26,466	18,738	66.3

Source Mission Kakatiya, Telangana State, 2021

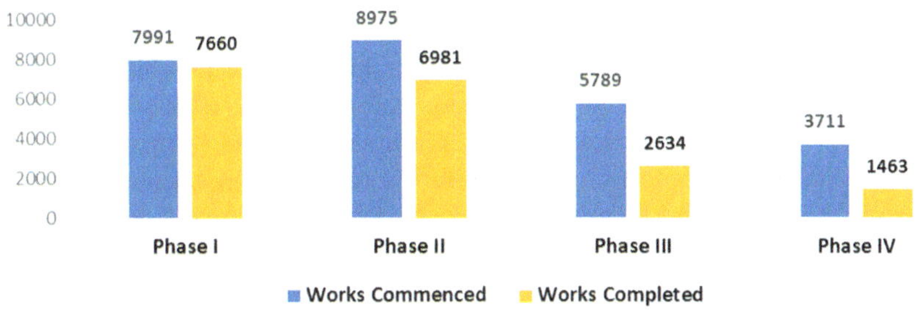

Fig. 12.1 Phase wise work progress of the Mission Kakatiya Programme (*Source* Mission Kakatiya, Telangana State, 2021)

43.8 lakhs hectares to 71.71 lakhs hectares. It is observed that implementation of Mission Kakatiya programme is one of the main factor in the increase of ground water levels and nearly 2.9 lakh of hectares of ayacut under the irrigation facilities. State government has introduced Rythubandu programme which ensures to the farmers to not fall in to the debt trap. The scheme is also called "Agriculture Investment Support Scheme" ("Rythu Bandhu"). It is also implemented across the state in the year 2018–2019 Karif season onwards. Initially they gave 4,000 per acre with no minimum and maximum land limitation and then in 2017 they increased 1,000 per acre with an Rs 5000/- per acre,

The socio-economic report (2019) reveals that, under the Rythu Bandu scheme, thc government identified a total 51.5 lakh farmers in the state and paid nearly 5,300 crore rupees during the 2018–2019 Karif season, whereas in Rabi season, 49.0 lakh farmers were benefitted. According to Socio-economic outlook—2021, the Government has transferred Rs. 14,651 crore rupees to 59.3 lakh farmers across the state. Later also the scheme covered almost Rs. 7,515 crores spent on the 61.5 lakh farmers.

12.6 Primary Data Results

Age group: 62% of the respondents are in the age group between 25 and 45 years and most of them are marginal and small farmers.

Awareness about the Government prestigious schemes: 93% of respondents have excellent knowledge of the prestigious government programmes. Because practically everyone exposed the programme, including Rythu Bandu, Rythu Bheema, Haritha Haram, Mission Bhagiratha and Mission Kakatiya. Among these, the two primary programmes that support farmers are Rythu Bandu and Mission Kakatiya.

Respondents view regarding Work progresss of Mission Kakatiya: In this regard, asked about the Mission Kakatiya works were commenced or not, in your residing villages. Out of 300 respondents, 96 of the respondents have mentioned that, the works were commenced and finished, only 4% respondents said, either works are not commenced or there is no tanks in the their villages. The respondents have given different responses on the benefit of the Mission Kakatiya programme.

View of respondents on Impact of Mission Kakatiya: 89% of respondents said Ground water levels have increased, created labour employment, siltation was done and there is impact on production of crop and 11% of the respondonents gave negative feedback and some of them not responded, stating that farmers didn't benefit from it and that contractors instead gained financially. But majority of the respondents mentioned that, the Mission Kakatiya is the good programme and it has given wonderful results in the state. According to them, after tanks rejuvenation works finished, the following benefits are received by the farmers and people. (1) The ground water levels increased in and around the tanks. (2) The siltation soil is given boosting to the high crop yielding, (3) Better thing is the prestigious scheme has created good employment during the drought and other situation.

Suggestions from respondents for the betterment of Mission Kakatiya Programme: Based on the primary data, total one third of the respondents have given suggestions on the betterment of the programme. A one fifth of the respondents said, government should focus on the depth of the tanks, which is not maintained properly in the digging process and also government should provide free distribution of soils to the needy farmers. 13% people are requesting to the government, along with tanks beautification and need to clean canal network and work also commenced by every year. Interestingly, 5% people requesting for, commencement of tank works under the Mission Kakatiya programme in their villages.

Possessing the agriculture land: 86% of the respondents have their own agricultural land and remaining 14% of the respondents they are leased farmers. If we look in to the acres wise, only 3% farmers own an average of 1acre land. The greatest (30%) percentage farmers have 1–3 acres land, and the remaining one fifth of respondents owning holdings between 3 and 5 acres. 4% of the respondents possessed more than 5 acres of land (Fig. 12.2).

Rythu Bandu scheme: Nearly three-fourth (86%) of the respondents said, they are receiving the Rythu Bandu, timely and rest of the respondents have said that they

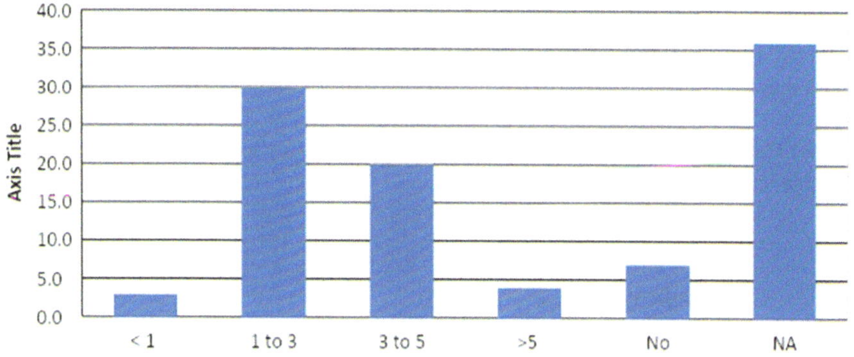

Fig. 12.2 Possessing the agriculture land (%) (*Source* Primary data)

are not receiving the agriculture investment support money, because of the assigned land or patta-related issues.

Suggestions from the respondents regarding Rythu bandhu: Nearly 60% of the respondents have suggested for the improvement of the farmer's development. Nearly one-fifth (22%) of the respondents suggested that, along with Rythu Bandu scheme, there is a need for other agriculture-based tools (fertilizers and tractors). 12% of respondents said that, the programme is good for all farmers, but there is a need to set up cutoff or maximum limit up to 5 acres land for the scheme eligibility. Only 10% of the respondents suggested that there should be increase in the investment amount up to Rs.10,000/- per acre. Few (5%) of the respondents mentioned that, the government is ignoring the tenant or leased farmers, who also need financial support from the government.

Very interestingly 10% suggested that, the Government has focused on the Agriculture and basic amenities which are very helpful to the needy people but the Govt also needs to concentrate on increasing health, education and employment services (Fig. 12.3).

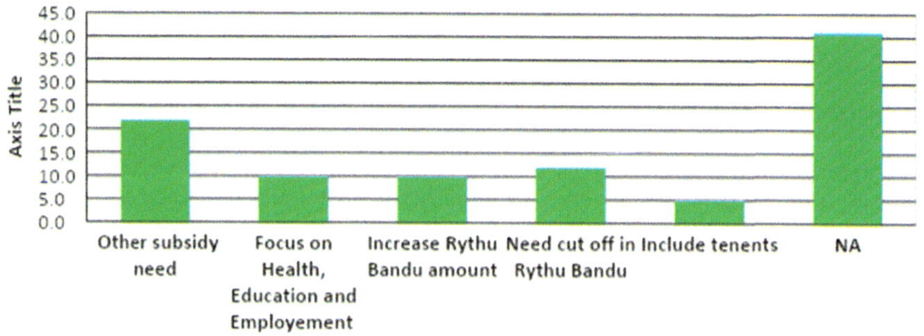

Fig. 12.3 Suggestions of respondents (*Source* Primary data)

The recent Central government statistics have shown there is a decrease in the pattern of farmer's suicide. With effective implementation of Rythu Bandhu scheme and other measures initiated for the farmers by the government, it is seen that farmers suicide in Telangana has a steep decline from 846 cases in 2017 to 491 cases in 2019. As per National Crime Bureau, Telangana state records the lowest number of farmer's suicide among major states.

12.7 Conclusion

Based on the previous experience, the state government has taken many good initiatives for the development of the farmers, in Telangana state. The study concluded that Mission Kakatiya programme has succeeded to rejuvenate the tanks, which is helping to increase the ground water levels and ayacut, across the State. Many states from the country have come to Telangana state and observed its implementation. Regarding the Rythu Bandu scheme, most of the community people like; Agency people, and tenant people who are having the forest land and actual cultivators are exempted from the prestigious Rythu Bandu Scheme. If government is implementing the scheme to tenant farmers, roughly 7.5–8.0 lakh tenant farmers will be benefitted. The socially and economically vulnerable people need support from the government to cope up their livelihoods. The primary data reveals that, majority of the farmers are requesting that, there is need to put cut off or maximum limit of 5 acres, which is imposed in the Central scheme, i.e., PM Kisan Samman Nidhi. It is also suggested that, above BPL farmers need to be exempted in this scheme. Government should initiate pension scheme for farmers those who lost assigned lands for the infrastructural development of the area.

References

Anjaneyulu et al (2021, March) Impact assessment of schemes introduced for drought prone districts of Telangana State. Int J Anal Exp Modal Anal XIII(III). ISSN No.: 0886 9367

India Today News paper. https://www.indiatoday.in/india/story/243-farmers-committed-suicide-in-telangana-govt-releases-compensation-of-rs-14-58-crore-1576481-2019-08-02. Accessed 3 Sept 2021

Office of the Registrar General & Census Commissioner, India, Ministry of Home Affairs, Government of India, Census of India. https://censusindia.gov.in/2011-common/censusdata2011.html. Accessed 12 Aug 2021

Official website of agriculture report. https://thewire.in/agriculture/ground-report-why-farmer-suicides-in-telangana-continue-unabated. Accessed 1 Sept 2021

Official website of Forest Department Government of Telangana. http://fmis.telangana.gov.in/Default_Forest.aspx,report. Accessed 11 July 2021

Official website of Government of Telangana. https://telangana.gov.in/PDFDocuments/Telangana-Socio-Economic-Outlook-2021.pdf

Socio Economic outlook of Telangana State—2021. Accessed 20 Aug 2021

Official website of Mission Bhagiratha, Government of Telangana, Status of Mission Bhagiratha on http://missionbhagiratha.telangana. Accessed 11 Aug 2022

Official website of Mission Kakatiya, Government of Telangana. https://missionkakatiya.cgg.gov.in/mis. Accessed 17 Aug 2021

Official website of NREGA Government of India. https://nrega.telangana.gov.in/Nregs/FrontServlet?requestType=KeyReportsRH&actionVal=Reports&page=KeyReportcenter_ajax_eng, State Level Consolidated Reports on www.nrega.telangana.gov.in. Accessed 3 July 2021

Chapter 13
Role of Women in Water Resource Management in Sikkim

Chhunu Maya Giri

Abstract Women have a strong affiliation with natural resources, as they are, by tradition, responsible for collecting fodder, grasses, fuel-wood and water and management of it for domestic purposes. Women across the world do play a vital role in natural resource conservation and management. This chapter analyses women's role in water management in water deficit and rain shadow zone of Darjeeling Himalaya. The findings of this study are based on intensive fieldwork conducted in five villages of Namchi, South Sikkim. The study reveals their overwhelming role in water collection but their near-complete exclusion from the decision-making process.

Keywords Water · Women · Water management · Participation

13.1 Introduction

In traditional economies, women have been responsible for collecting fodder, fuel-wood and fetching water and management of it for domestic purposes. Women across the world have been playing a vital role in natural resource conservation and management. Fuel, fodder and water collections were the conventional responsibilities of women (Sharma and Kaushik 2011).

Since the 1970s women in several parts of the world have started organizing programmes to stop the degradation of water resources and other natural resources. One such movement is Green Belt Movement initiated by Nobel Prize winner Wangari Maathai in 1977 for planting trees. Similarly, Maria Cherkasova a journalist, ecologist and director of the Centre for Independent Ecological Programs (CIEP) worked for several years to stop the construction of a hydro project dam in Kutan River (Aditya 2016). Women of Lima, Indonesia formed a group and

C. M. Giri (✉)
Department of Geography, Sikkim Government College Rhenock, Sikkim University, Gangtok, Sikkim, India
e-mail: chhunugiri22@gmail.com

A. Dubey et al. (eds.), *Promoting Sustainability Through Water Management and Climate Change Adaptation*, Advances in Geographical and Environmental Sciences,
https://doi.org/10.1007/978-981-99-5479-7_13

started a water project. At first, they chose the locations and then raised the initial capital by cultivating a community field and later invited men into the local water committee (Resource Guide, 2006). The role of women in environment protection in India can be traced back to 1731 AD when Amrita Rai sacrificed her life to save the trees of her village. Yet another movement the Narmada BachaoAndolan (Save Narmada Movement) has been led by MedhaPatkar protested for years to render support to thousands of displaced villagers due to submergence owing to the dammedriver (Bandyopadhyay 1999).

It is primarily women who bear the daily burden of hauling heavy buckets of water over long distances to meet the domestic water needs of their families (Ahmed 2006). Village women are considered privileged water managers in the concepts of 'eco-feminism'. It tends to position women as natural caretakers of the environment because of their childbearing and nurturing roles. Women are responsible for collecting water for domestic needs such as drinking, cooking, cleaning and washing work that is unpaid and time-consuming (Ahmed 2006).

There are several pieces of evidence documented on the increasing role of women in water resource management. For instance (Aladuwaka and Janet 2010), has highlighted the innovative women-run water project of WanaraniyaMatale District of Sri Lanka, where women collectively formed Vishaka Women Society and had initiated a water project across the perennial stream of Bambara Kiri Ella, 6.5 km away from the village. Its main objective was to solve the water problem of the village. It was a women's society that successfully carried water through a pipeline to every household in the village. Efforts of the women not only provided safe drinking water to the villagers but also helped in income generation.

Likewise (Were et al. 2006), in their study in western Kenya found that women had formed women groups for the management of water projects when they felt marginalized in the water project. They raised funds by engaging as labor in harvesting and weeding. Money obtained was invested in planting tea seedlings which were later sold. Some amount of earnings they distributed amongst the member and the left amount each member contributed for repaying loans which were obtained by men for the construction of the reservoir tank. Similarly in another village in the same district, women's groups have contributed from the monthly earning of Rs. 500 toward the spring protection and purchase of pipes.

In Gujarat, Self Employed Women Association (SEWA) educated and trained around 2 lakh rural women of 500 villages of the semi-arid region of Gujarat on the method of water harvesting and awareness generation. They are involved actively in water campaign awareness activities, such as rainwater harvesting at the household and community level, the revival of traditional water sources and watershed development (Prashad, n.d.).

These are the various ways women have participated in natural resource conservation and management. Women are always close to natural resources like forests and water because most rural women are daily involved in the collection of fodder, firewood and water for their household. The depletion and degradation of any of these resources affect women the most.

On the other hand, related studies highlighted how social structure and composition brought a gender disparity in the water management sector. Still, women largely were not been involved in the water management sector and their role was considered invisible. For instance (Ruth and Dick 1998), argued that the trend of transferring irrigation management responsibility from the state to communities or local user groups has ignored the implication of intra-community power differences, and gender differences for the effectiveness and equity of water management.

Similarly, the studies of Prakash and Sama (2006) and Joshi (2011) in the Kumaon region and Gujarat respectively have found that gender disparity in water access was high and women from Dalit family were more vulnerable during the dry season because they were restricted to access water from upper cast people as they owned the water sources. Upper caste people were more benefited by schemes and projects as compared to a lower caste (Khandelwal et al. 2012) did a comparative study of the involvement of tribal and non-tribal women in water management in Rajasthan from four villages with 200 respondents and found that the women involvement of tribal women in the planning stage is very low.

The existing literature explored that less research has been done on women's participation in water professional in South Asia, especially in India (Kulkarni 2013). There should be a change in the existing understanding of women's work to make a change in respect to women's representation in the water sector. Some suggestions which have been put forward in this study are gender policy for water bureaucracy, gender-specific benefits and amenities, women-specific training, etc., to increase the presence of women in the water sector (Satya and Cronin 2013).

There have not been much studies on the role of women and their participation in the management of water resources in Sikkim. Most of the studies in Sikkim have mainly focused on climate change and its impacts. Like in other parts of the world, women in Sikkim have been playing a vital role in water management both at the household level and in society. This work becomes more significant with the fact that Namchi has traditionally been a rain shadow zone of Darjeeling Himalaya.

13.2 Data and Methods

The objective of the present study is to understand the participation of women in water resource management in Namchi and to examine gender gap in water fetching. Hence, the present study is based on the primary data collected through field survey conducted in five villages of Namchi. The selection of sample village was done by applying purposive sampling technique. A total of 180 households from five villages has been selected as sample for the study. Primary data has been collected through interview, Focused Group Discussion (FGD) and observation methods (Table 13.1).

Table 13.1 Sample size

SL. No	Name of the purposive selected GPU	Purposive selected villages	No. of selected households for survey
1	Rangbul	Rong	33
2	ManiramPhalidara	Lower Phalidara	39
3	KitamManpur	Upper Kitam Ward	38
4	SorokShyampani	Upper Sorok	30
5	Mikkhola	Singithang	40
	Total		180

13.3 Study Area

Namchi is one of the revenue blocks of the South district of Sikkim. It covers the total geographical area of 552.93 square kilometre of South Sikkim, which constitutes 73.72% of the total land area of the South District. In Namchi rural areas covers 545.33 square kilometre and the urban area covers only 7.60 square kilometre, which accounts for 98.62 and 1.37% respectively of the total area of Namchi.

The area comprises steep slopes and rugged topography. The elevation of Namchi varies from 193 to 2643 m from mean sea level (MSL). The general slope is from the North toward the South. Namchi receives almost one-third of the rainfall received by Gangtok, which receives almost 3250 mm of annual rainfall (Sharma 2013). The governing factor for this may be its location, as it is located in the rain shadow zone of Darjeeling Himalaya (Map 13.1).

13.4 Results and Discussion

Women and young children are mostly seen collecting water for their household requirements. Figure 13.1 demonstrates that out of 98 water fetching households, females from 70% of households have participated in water collection whereas, males from only 28% of households, this shows that the burden of water collection falls disproportionately on women. The participation of males in water collection is very less across the country. The data collected from IIM (India Institute of Management) Delhi 1998 for Rajiv Gandhi Rural Drinking Water Mission shows that only 6% of males fetch water.

It was observed during the field survey that in most of the household females were carrying water in pots and oil jars (up to 50 L of water in a single trip). Further, it was observed that females were more interested and confident to respond than males regarding water-related problems and their use of it at the household level.

The content in the box below reflects the perspective of the respondent regarding the women and water collection (My neighbour collects more water than us because

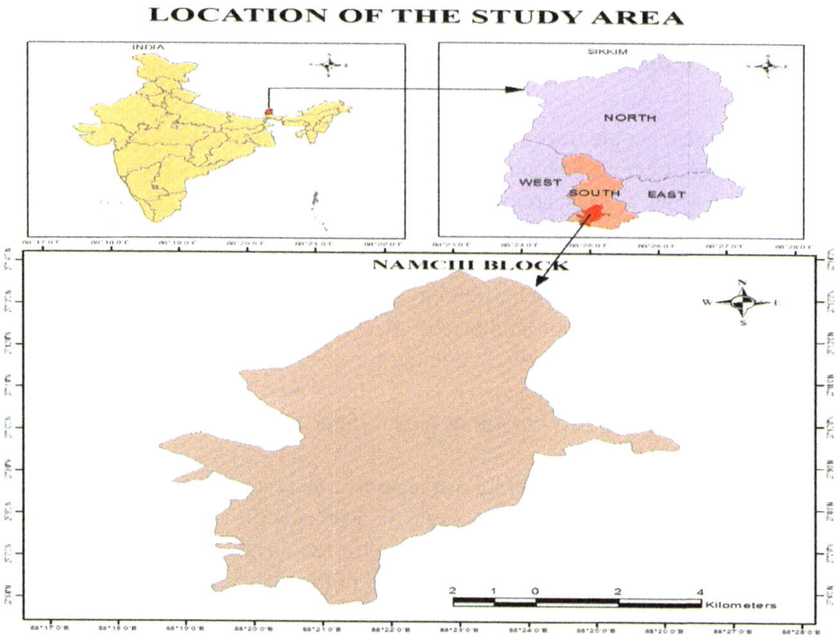

Map 13.1 Location of the study area (*Source* Modified after RMDD, Government of Sikkim, Projected-UTM WGS 1984-Zone 45N)

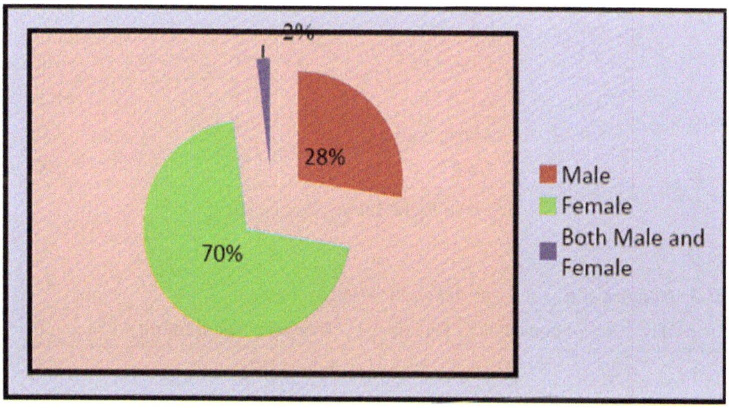

Fig. 13.1 Proportion of male and female participation in water collection (*Source* Field Survey 2015–2016)

they are having a large number of female members to collect water). Deepa Gadaily, Upper Sorok.

13.4.1 Distribution of Households by Distance Covered for Collection of Water

In the study area, people travelled a different distance to collect water to fulfil their daily water requirements. Moreover, in the lean period (from November to April) people faced acute water scarcity, as the seasonal streams dried up and also these months recorded almost no rainfall. Because of this reason more than three-fifths of households travel up to half a kilometre distance to collect water daily for several trips. Similarly, around 34% of households covered up to 1 km distance, two to three trips daily. Even a significant number of households travelled 2–3 km distance to fetch water for daily consumption. Women travel early morning before they start their household work for such a distance to collect water. I can get chance to collect water only if I wake up early morning 4 Am. (Statement by Sangita Subba, Upper Kitam village) (Table 13.2).

It is not only that people travel far distances to collect water but they have to cover this distance twice and thrice a day. Table 13.3 reflects the number of trips people make to collect water and the total time taken for such trips. Firstly, data on several trips of water collected per day reveals that half of the surveyed households make up two trips to collect sufficient water per day.

Table 13.2 Distribution of households by distance covered for collection of water

Distance in (Kilometre)	No. of household	Percentage
Below 0.5 km	59	60.20
0.6–1 km	22	22.45
1.1–2 km	10	10.20
2.1–3 km	6	6.12
Above 3 km	1	1.02
Total	98	100

Source Field Survey, 2015–2016

Table 13.3 Number of trip and time taken for water collection

No. of trips/day	No. of household	Percentage	Total time taken for total trip	In Percentage
1	25	25.51	Below ½ hr	32.29
2	49	50.00	1/2–1 hr	34.37
3	13	13.26	1–1.1/2 hr	6.25
4	5	5.10	1.1/2–2 hr	11.45
5	3	3.06	2–2.1/2 hr	3.12
7	2	1.04	2.1/2–3 hr	6.25
14	1	1.02	Above 3 hr	6
Total	98	100.0		100.0

Source Field Survey, 2015–2016

Further, around 23% of households make 3–7 trips to bring water for their daily consumption. Hence, engaging oneself from the productive time for a whole day to collect water is not only an economic loss but also physical stress. A close look at the data indicates that on average 67% of households spend 1 hr on water collection. Similarly, a significant number of households spent up to 3 hr and one household spent more than 3 hr for 14 trips per day to collect water. From this household, two members collect water 7 trips each so altogether they do 14 trips. It was revealed during the interaction with the people that during the dry period the task of one or two people especially the women from each of the households from Upper Sorok village to collect water the whole day.

13.4.2 Community Participation in Water Management in Rural Area of Namchi

The analysis of the gender gap in participation reveals that male participation is higher than the participation of females in water management aspects. On average males from 84% of the household and females from only 16% of households participate in water management. The reason for the high number of male participants is that the locations of water sources are very far from the village and have to go through rugged topography and cliff which requires rigorous physical work. The study reveals that 16% of the female are mostly participating in the plantation and cleaning of the source area.

There are different water management committees formed in the study area viz Water Harvesting Committee, Village Water and Sanitation Committee and Water User Committee. These committees are formed by the villagers for the management and distribution of water within the village and committees are functioning through a self-financed system (Table 13.4).

Although the water management committees of the villages are functioning smoothly yet there exists gender disparity in terms of membership. Figure 13.2 reveals that males, from 82% of households are the members of committee whereas females from only 17.8% of households are the members of the committee. Further, it

Table 13.4 Water management committees formed in rural Namchi

Name of water management committee	No. of HH	In percentage
Water Harvesting Committee	10	22.2
Water User Committee	28	62.2
Village Water and Sanitation committee (VWSC)	6	13.3
Jal Kranti Committee and VWSC	1	2.2
Total	45	100.0

Source Field Survey, 2015–2016

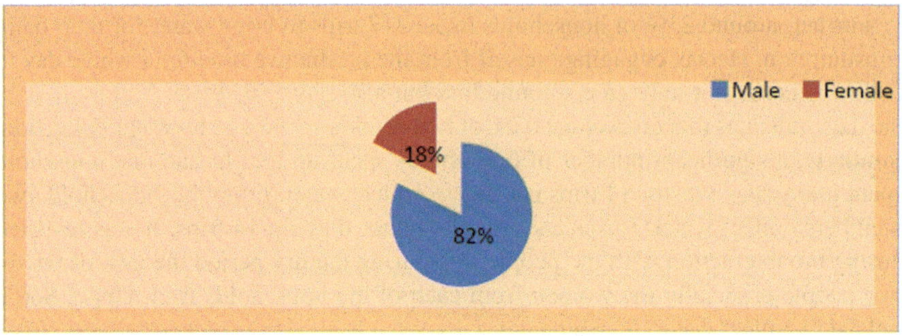

Fig. 13.2 Percentage distribution of male and female participation in water management committee (*Source* Field Survey, 2015–2016)

was found while interacting with the women during the fieldwork their involvement in the annual meeting is very low. This reflects that the participation of women is very low in the water management committee. This signifies that women are nearly excluded from the committee. But in terms of water collection, their role is overwhelming. Many studies on women's participation in water resource management found very low participation of women in water management sectors for instance the UNESCO New Delhi Report on Water User Association for sustainable water management in Tamil Nadu (Kulkarni 2013). This is relevant to Namchi also as the participation of women is very low in the water management committees.

13.5 Conclusion

Globally, women are prime users and protectors of water. They are playing a significant role as water collectors or managers within a household. It is women who do much of the collection of water and other natural resources to sustain their families. The finding of the study reflects that women are playing a vital role in water resource management for domestic purposes. As the study reveals that from the surveyed household 84% are women who collect water for their household requirements and manage it within the house for different purposes. On the other hand, their participation in the construction and maintenance of pipelines is low as in most of the household male members are participating in such work which is mainly due to the far location of sources. It is very surprising to know that the proportion of women's participation in different management committees is very low which signifies that they are nearly excluded from the decision-making process.

Women's resource management work should not be overlooked they should get equal participation in all aspects of water management. They should not only consider water collectors. There should be a provision for equal women's seats in water management committees and should be involved in the decision-making process as well.

Acknowledgements Author would like to express sincere gratitude to the people of Namchi Block of South Sikkim who sincerely responded to her questionnaire and provided first-hand information. Author also thank supervisor Prof. Sohel Firdos, Department of Geography, Dean of Human Science, Sikkim University for guiding me in writing this chapter. Author also expresses gratitude to Dr. Uttam Lal, Assistant Professor, Sikkim University for encouraging and helping me to complete this chapter. My sincere thanks go to Dr. Anupama Dubey Mohanty, Assistant Professor, Institute of Public Enterprise, OU Campus Hyderabad for consistent response in the process of selection of my paper for publication.

References

Aditya SK (2016) Role of women in environmental conservation. Int J Polit Sci Dev 4(4):140–145

Ahmed S (2006) Mainstreaming gender equity in water management: policy and practice. In: Reddy VR, Dev SM (eds) Managing water resources: policies, institutions and technologies. Oxford University Press, New Delhi, pp 249–264

Aladuwaka S, Janet M (2010) Sustainable development, water resource management and women's empowerment: the Wanaraniya water project in Sri Lanka. Gend Dev 18(1):43–58

Bandyopadhyay J (1999) Chipko movement: of floated myths and flouted realities. Econ Polit Wkly 34(15):880–882. https://www.jstor.org/stable/4407841

Joshi D (2011) Caste, gender and the rhetoric of reform in India's drinking water sectors. In Shah M, Vijayshankar SP (eds) Water: growing understanding, emerging perspectives. Orient Blackswan Private Ltv., New Delhi

Kulkarni S (2013) Situation analyses of women water professionals in South Asia. South Asia Water Stud 3(3):73–91

Khandelwal N, Sharma VP, Sharma C (2012) Involvement of women in natural resource conservation in Udaipur district of Rajasthan. Raj J Extn Edu 20:224–228

Prakash A, Sama RK (2006) Social undercurrents in a water-scarce village. Econ Pol Wkly 1(7):577–579

Prashad (n.d.) Women and environmental movements and policies in India paper—9 module 24, 5–7

Resource Guide (2006). Mainstraming gender in water management. Version 2.1, November 2006. Available online at https://www.undp.org/site/g/files/zskgke326/files/publication/IWRMGenderResourceGuide

Ruth M, Dick MZ (1998) Gendered participation in water management: issues and illustrations from water user association in South Asia. Agric Hum Values 15(14):337–345

Satya P, Cronin A (2013) Water in India; situation and prospects. UNICEF, FAO and SACI Waters

Sharma S (2013) Climate change indicators in Sikkim with reference to Gangtok and an emphasis on Hydrology. PANDA—ENVIS Sikkim Newsl 5(1):32–34

Sharma R, Kaushik DB (2011) Role of women in environmental conservation. Int J Multi Manage Stud 1(2):162–167

Were E, Swallow B, Roy J (2006) Water, women and local social organization in the Western Kenya highlands, ICRAF Working Paper no. 12, Nairobi, World Agroforestry Centre, p 36. Available online at www.capri.cgiar.org

Chapter 14
Assessing Human-Wildlife Conflicts in Tiger Corridor Habitat: A Case Study of Ranthambhore Tiger Reserve, India

Bhanwar Vishvendra Raj Singh, Anjan Sen, Manjit Singh, Ritika Prasad, and Ravi Mishra

Abstract Human-wildlife conflict is a very common phenomenon in any protected area. Man-environment relationship research has been conducted on human-wildlife conflicts in different locations of the world. However, the nature and extent of human-wildlife conflicts vary from place to place. This paper examines human-wildlife conflict activities by animals in the Ranthambore Tiger Reserve (RTR) on various livestock, using data collected over sixteen years (2001–2016). Research data is collected through qualitative information such as questioning, observation, field-work, personal interviews, and other methods. Quantitative information has also been collected from RTR Forest Office. One thousand five hundred and forty-six were killed along with tigers, leopards, bears, and other wild animals. Most of the livestock killings were of domestic cattle such as cows, bulls, buffaloes, goats, and others. The leopard is an important wild animal that was often targeted at cattle. After that, the tiger is the second predator. Overall, livestock hunting was highest in summer and low in other seasons. Phalodi and Khandar districts were identified as the most 'poaching hotspots' with a high proportion of deaths in the relative abundance

B. V. R. Singh (✉)
Department of Geography, Faculty of Earth Sciences, Mohanlal Sukhadia University, Udaipur 313001, India
e-mail: bhanwarsa28@mlsu.ac.in

A. Sen
Department of Geography, Delhi School of Economics, University of Delhi, Delhi 110007, India
e-mail: asen@geography.du.ac.in

M. Singh
Department of Geography, School of Earth, Biological and Environmental Sciences, Central University of South Bihar, Gaya 824236, India
e-mail: manjitsingh@cusb.ac.in

R. Prasad
Department of Geography, University of Lucknow, Lucknow 226007, India

R. Mishra
St. Department of Remote sensing and GIS, Kumaun University, Almora 263001, India

A. Dubey et al. (eds.), *Promoting Sustainability Through Water Management and Climate Change Adaptation*, Advances in Geographical and Environmental Sciences,
https://doi.org/10.1007/978-981-99-5479-7_14

of livestock. The proximity of core areas and the implications of our findings for reducing livestock loss in RTR and conservation of large carnivores are discussed.

Keywords Human-Wildlife conflict · Livestock · Carnivores · Poaching hotspots · Tiger habitat management

14.1 Introduction

In the contemporary world, human-wildlife conflict is a rapidly increasing phenomenon near the surrounding protected areas. It's become a big challenge in the man-environment relationship in the world (Thirgood et al. 2005). Crop raiding, property damage, livestock depredation, and human casualties are the most varieties of clashes with wildlife (Ogra and Badola 2008; Inskip and Zimmermann 2009). Human casualties and livestock depredation are the gravest nature of the conflict between all (Baraso 2012).

The man-wildlife conflicts arise once they are forced to share a standard limited resource such as land, game, livestock, or fish (Sillero-Zubiri and Laurenson 2001; Schwerdtner and Gruber 2007). Also, the following are supposed to be the effect of conflicts between humans and wildlife. The impacts of transportation and animals occur when aircraft are struck by bird's hits and when humans are bitten by contaminated wildlife (Messmer 2000; Conover 2001). Mostly, all human-caused mortalities of wild animals, including poaching for the trade of body parts, will be considered as human-wildlife conflict (Everatt et al. 2019).

The critical variety of human-wildlife conflict is the loss of human life (Thirgood et al. 2005; Gurung et al. 2008) and therefore the retaliation against wild animals. Such reciprocal persecution in defense of livestock and protection of crops threatens the survival of this wildlife that comes into conflict (Mishra 2001; Mishra et al. 2003).

Continuously, many carnivores have extinct, for example, the Javan tiger and the Bali tiger (Nyhus and Tilson 2004). The population of various species like the Asiatic lion (Panthera leo), the Cheetah (Aconytusjubatus), and therefore the tiger (Panthera Tigris) has declined significantly due to the human-wildlife clashes (Treves and Karanth 2003).

Human-wildlife conflicts are going to be of specific concern once they are related to big carnivores attacking humans like tigers in Asia and lions in Africa (Dickman and Hazzah 2016). When the people involved are poor, damage caused by wildlife may significantly affect the livelihood of local communities. If the predators involved in conflicts are of legally protected status or are endangered, the consequences of conflict can be controversial (Redpath et al. 2013; Graham et al. 2005). If damages relentlessly affect the livelihood of native communities, getting their active support, which is indispensable for conservation can be difficult (Mishra 1997).

The wildlife-human conflicts bring much social, economic, and ecological significance. People frequently migrate from wildlife-conflict areas to zero-conflict areas (Gilleland 2010). Harvest and property damage and livestock depredation are public effects resulting in massive economic losses worldwide (Thirgood et al. 2005). These effects undermine the administrative support for conservation in protected areas and should need the extinction of the problem animal (Treves and Karanth 2003).

For illustration, the rapid success of community forestry agendas nationwide and the initiation of conservation of buffer zone forests neighboring the parks and reserves after the 1990s formed (Chandy and Euler 1998). The other habitat beyond the national parks and reserves and therefore the consequential movement of wildlife in these freshly developed habitats resulted in an amplified frequency of human-wildlife conflict (Gurung et al. 2008).

Much research has been conducted on human-wildlife conflicts in different locations of the world. However, the nature and extent of human-wildlife conflicts are different from place to place (Sillero-Zubiri et al. 2007). The conflict mitigation measures applied in one locality may not fit well in other areas. Because socio-political, cultural, economic, and geographic situations are not the same for all places (Adger 2006). Thus, the principle of one-size-fits-all cannot be applied universally (Msoroka and Amundsen 2018).

14.2 Study Area

Ranthambhore Tiger Reserve is located between latitudes 25,052′071″ N to 260 33′713″ N to Longitudes 75,085′84.0″ E to 77,002′48.0″ E at the junction (Great boundary fault) of Aravallis and Vindhyan ranges (Fig. 14.1).

This core area is 1113.36 km^2 and the buffer area is 297.92 km^2. It has a total geographical area of 1411.28 km^2. It is the single largest expanse of dry deciduous Anogyssus pendula forest that is intact in India. At present, the total number of tigers within the Ranthambore Tiger Reserve is above seventy. Through its conservation, there are so many land ecosystem services provided to stakeholders through tiger conservation as well as Ranthambore is the foremost hotspot tiger reserve and marks the transition zone between the true deserts and seasonally wet peninsular India.

14.3 Data and Methods

14.3.1 Database

Research data is collected from both qualitative and quantitative sources, qualitative information was gathered through questioner, observation, fieldwork, personal interview, and other skills. Quantitative information has been collected from several organizations like Census of India 2011, Ministry of Environment and Forests and Climate Change (MoEFCC), National Tiger Conservation Authority (NTAC), National Atlas, and Thematic, established under Wildlife (Protection) Amendment Act 1972. Indian

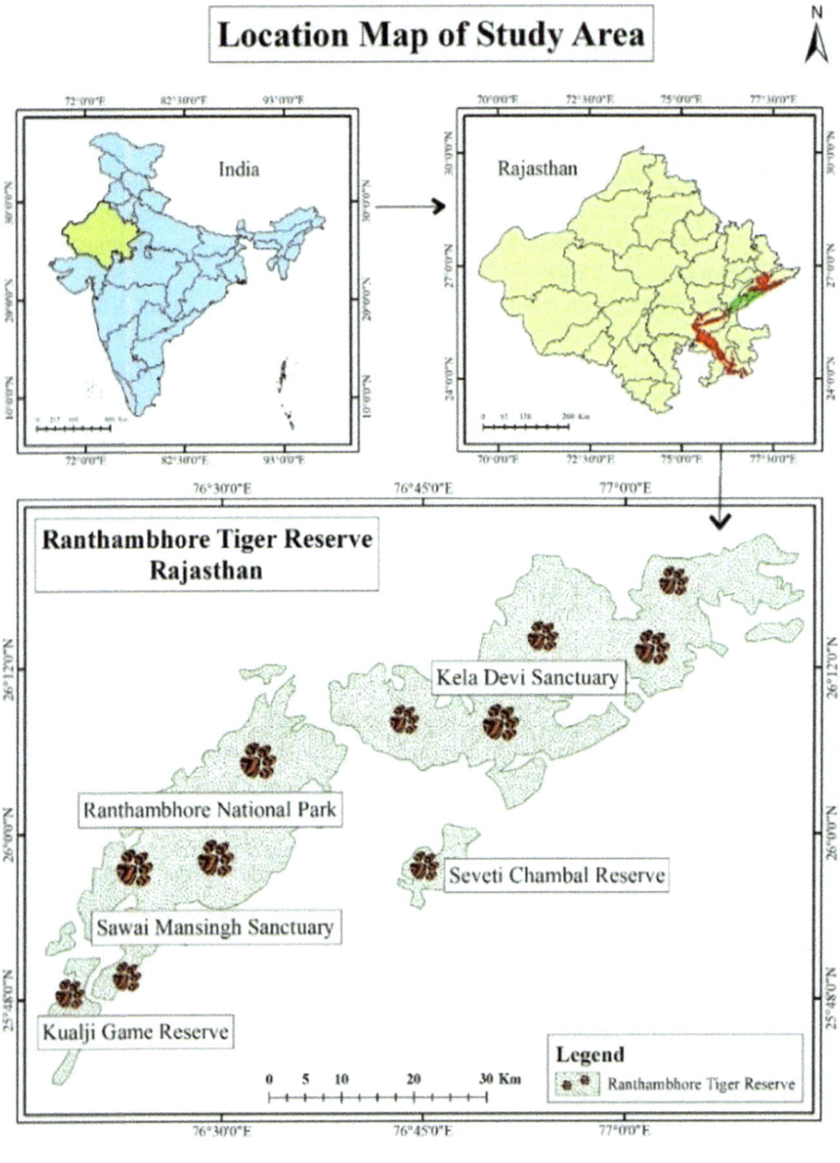

Fig. 14.1 Study area of Ranthambhore Tiger Reserve, India

Remote Sensing Satellites available with Mapping Organization (NATMO), National Remote Sensing Agency (NRSA), Hyderabad, Forest Department (Government of Rajasthan), Ranthambore Forest Department, and other NGOs.

14.3.2 Data Analysis

The qualitative data has been collected and scrutinized for data study. In this method, first of all, preparation for base map through digitization of NATMO districts boundary with reserve forest areas in two districts in Rajasthan, where reserve lies. Maps have been done by calibration of Georeferencing and mosaic methods. Finally, prepare a base plate which is the most essential part of data analysis. Further, demarcations of the study area and the village-wise boundaries have been created based on the village directory, 2011.

After that, the human-wildlife conflict data has been attached to the digitized map and raster maps are created using those attributes. The conflict indicators are collected in vector formats. Using all the vector maps, the vulnerability of the reserve has been done through spatial analysis using the vector calculator tool in Arc GIS, ERDAS IMAGINE, Microsoft Excel 2010, and SPSS 15.0. Linkert scale has also been used to quantify the perception of respondents.

14.4 Result and Discussion

This research provides insight into the present scenario of human-tiger conflicts in a very lowland protected area in the Ranthambhore Tiger reserve and recommends conservation and management strategies to decrease man-tiger conflict. The area of study is densely inhabited by humans in a boundary setting who depend significantly on forest resources like timber, fuelwood, fodder, grasses, and different other non-timber jungle products and herbs as food and for use in traditional treatments (Nasi et al. 2008).

Human-wild animal interactions often influence carnivore conservation and result in mitigating encounters (Sillero-Zubiri et al. 2007). We studied man-tiger (Panthera tigris) conflicts in pastoral villages near Ranthambhore Tiger Reserve (RTR), Rajasthan, India for six years (2005–2011) and categorized and studied the causes of conflicts. We documented 113 human-tiger conflicts. Most of the encounters between humans and tigers were from attacks on domestic livestock (88.5%) and humans (11.5%). Among livestock, cows, oxen, and calves accounted for 31.6, 21.1, and 16.7%, respectively, followed by buffaloes (19.3%) and goats (11.4%) after the death of tigers. Livestock looting was found within villages (53.4%), agricultural areas (44.5%), and forests (1.9%). We recorded 13 attacks on humans: nine were not fatal, but four resulted in death.

The cluster of forests comprising the Ranthambhore Tiger Park is surrounded by densely populated villages that depend on jungle resources for their livelihood and energy needs. This ends up in infrequent cases of human-wildlife conflict, where villagers are injured in encounters with wildcats like tigers and leopards when they enter the forest for the collection of firewood and fodder. Each year, numerous cases of cattle lifting are reported by villages, as tigers and leopards stray into the human residences in search of easy prey.

14.4.1 *Human-Wildlife Conflict in Ranthambhore Tiger Reserve, 2001–2016*

Figure 14.2 shows the entire seasonal human-wildlife conflict from 2001 to 2016 within the Ranthambhore Tiger Reserve. It shows a stable conflict rate between 2001 and 2007 followed by a sudden rise. Then after a rapid rise in the rate of conflicts due to numerous cases of the conflicts and other issues are noticed for the sudden rise within the rate of conflict.

There are so many reasons for increasing conflicts as a continuous tiger population is increasing from the beginning, absence of fencing of core and buffer areas, unfair human activities, scarcity of water, food and shelter is the main cause, etc., throughout things have changed human as a way more careless, rising violence of animal due to one-sided human activities, rising environmental imbalance (Fig. 14.2).

Dropping tolerance capacity of a human, in this phase saw a sudden rise in between the period of 2007 and 2012, then after that there is a drastic downfall in the conflict rate by providing protection measures through fencing, creating awareness about the importance of ecological balance, effective role of management authority, better monitoring of tiger movement as well as another wild animal.

It shows the year-wise total seasonal human-wildlife conflicts in Ranthambhore Tiger Reserve from 2001 to 2016. According to the graph from 2001 to 2006, there is conflict rate constant after that we can notice a sharp upsurge in the number of conflicts, from 2007 to 2012 and frequency is too much high, which has given the

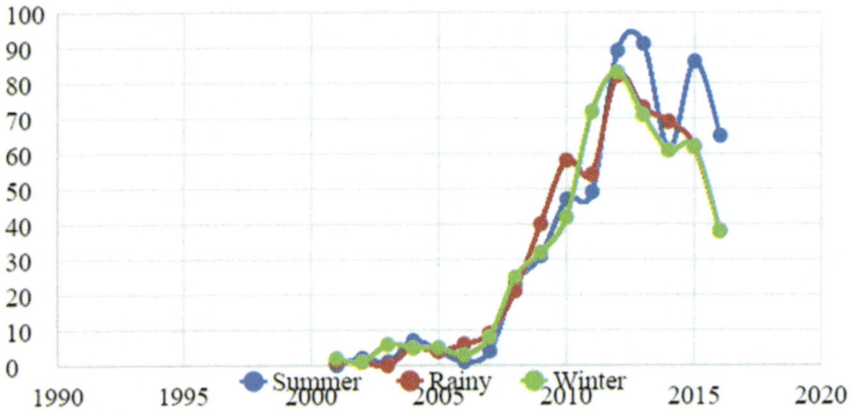

Fig. 14.2 Year-wise seasonal human-wildlife conflicts in Ranthambhore Tiger Reserve, 2001–2016 (*Source* Forest Department, RTR, 2016)

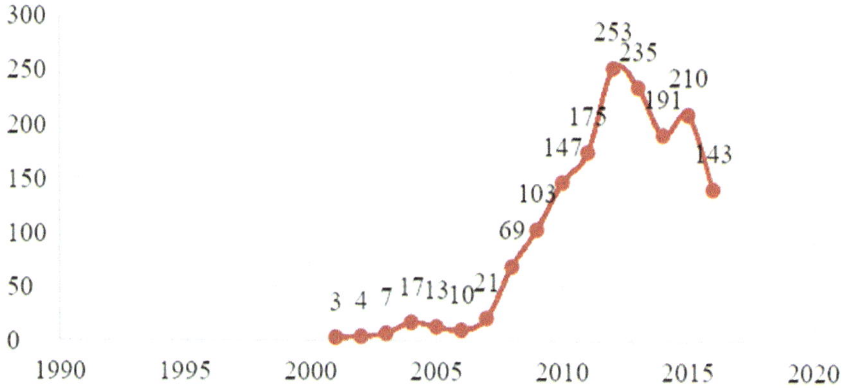

Fig. 14.3 Year-wise total human-wildlife conflicts in Ranthambhore Tiger Reserve, 2001–2016 (*Source* Forest Department, RTR, 2016)

maximum number of human-wildlife conflicts, the maximum number is 253 conflicts in the year 2012 (Fig. 14.3).

Therefore, there are many factors as, degradation of a human moral natural imbalance, abnormal shortage of resources due to excess harnessing of resources, pastoral movement by the herders, movement in search of water by wild and domestic animals from 2007 to 2012.

Then after 2012 a decrease in the conflict frequency, the primary period people were not so well awaked in Ranthambhore Tiger Reserve. After that local person is an awareness of the conservation program with proper utilization of natural resources, the rate of development was nominal, slow pace of urbanization, deforestation at a low rate, less destructive activities by the human, normal scarcity of resources in near Ranthambhore Tiger Reserve.

The graph shows year-wise total human-wildlife conflicts in Ranthambhore Tiger Reserve, (2001–2016). The graph is divided into four categories as a Tiger, Leopard, Bear, and Others wild animals show how much tiger, leopard, bear, and other wild animals (Wild Boars, Hyenas, Fox's crocodiles, etc.) kill per year. We find three phases, first phase is showing constant human-wildlife conflicts which began from 2001 to 2005 there were very few cases that happened but after 2006 to 2012 human-wildlife conflicts were growing continuously. But again from 2014 to 2016 decline phase was started which is a very positive sign to reduce the conflict rate in Ranthambhore Tiger Reserve and also local stakeholders. In this middle period, HWC highly registered case because of tiger and leopard population is increasing and also livestock increasing. In that condition, there are so many requirements for both sides (Fig. 14.4).

According to the graph, it's showing total human-wildlife conflicts in the Ranthambhore Tiger Reserve from 2001 to 2016. It is divided into four categories as a tiger, leopard, bear, and other wild animals, and every category shows how much conflict with domestic cattle's as cows/oxen, buffaloes, goats, and other domestic

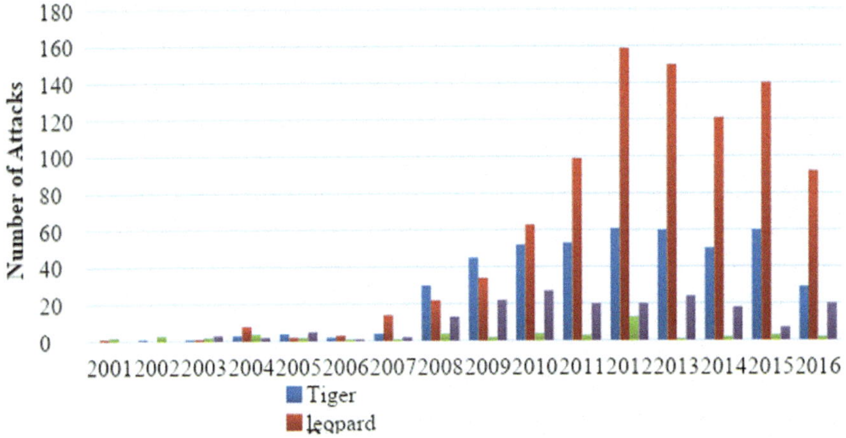

Fig. 14.4 Year-wise total attacks by wild animals in Ranthambhore Tiger Reserve, 2001–2016 (*Source* Forest Department, RTR, 2016)

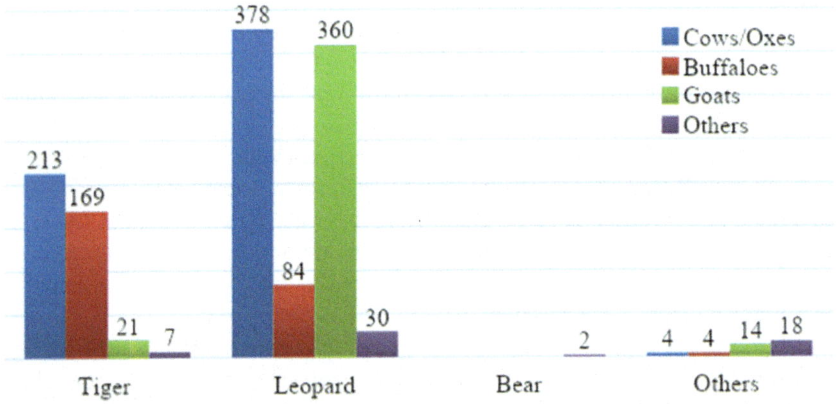

Fig. 14.5 Total domestic animals hunted by wild animal's attacks in Ranthambhore Tiger Reserve, 2001–2016 (*Source* Forest Department, RTR, 2016)

cattle. And it can say how much tiger, leopard, bear, and other wild animals (Wild Boars, Hyenas, Foxes, Crocodiles, etc.) are hunted. The graph shows two wild animals are mainly dominating on the conflicts, first one leopard and the second one is Tiger. Leopard is a crucial wild animal that frequently targets cattle, with the prime choice being cow/oxen and goats resulting in the highest number of conflicts with a leopard in proportion to other wild animals. After that, the tiger is the second predator targeting especially cow/oxen (Fig. 14.5).

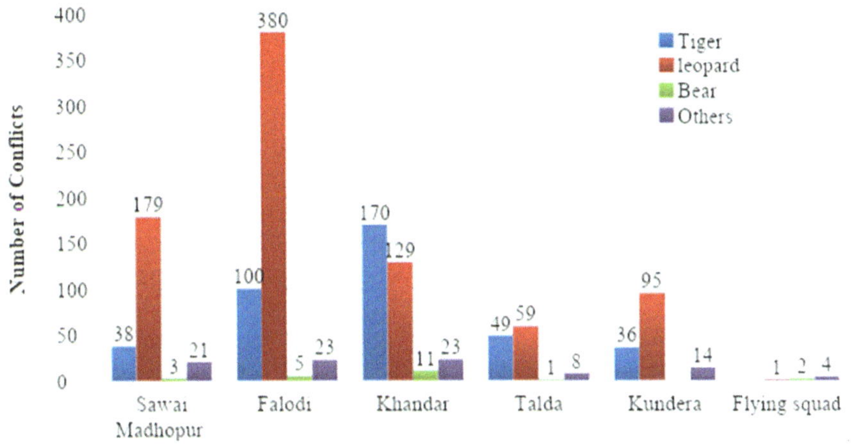

Fig. 14.6 Range-wise human-wildlife conflict in SWM Division, 2001–2016 (*Source* Forest Department, RTR, 2016)

14.4.2 Range-Wise Human-Wildlife Conflict in SWM Division, 2001–2016

This graph is showing the range of wise human-wild animal conflict in the SWM Division from 2001 to 2016. According to the graph, in the Falodi range, the highest 380 cases happened by leopard attacks. As follows Sawai Madhopur and others range. Tiger wild animal, in the Khandar range 170 cases happened due to the strategic location of this town and in the Falodi range 100 cases happened also.

Other wild animals and bears are also involved in conflicts. Khandar, SWM, falodi, and kundera ranges have seen some cases registered which are vulnerable to other wild animals also. Bear attacks are dominant in Khandar ranges given the location of the town, proximity to the forest boundary, and water scarcity is resulting in human-wildlife conflicts (Fig. 14.6).

14.4.3 Range-Wise Human-Wildlife Conflict in Karauli Division, 2001–2016

This graph is showing the range of wise man-wild animal conflicts in the Karauli Division from 2001 to 2016. According to the graph, In the Baler range, the highest 27 cases happened by tiger attacks. As follows Sapotra and others range. Leopard wild animal, in the Karanpur range 21 cases happened due to the strategic location of this town and in the Sapotra range 11 cases happened.

Kaila Devi, Karanpur Sapotra has the highest number of cases of conflict with a bear in comparison to attacks in RNP. Other wild animals (Wild boar, fox, hyena, and crocodile) proved more lethal in the Baler range where 56 casualties were reported—the highest caused by beasts other than tiger, leopard, or bear. Few cases were also reported in Mandrail and karanpur. This reflects that in the Baler range the concentration of other wild animals is higher than the tiger, leopard, or bear.

Especially, the Baler range in the Kaila Devi wildlife sanctuary is highly vulnerable to human-wild conflicts due to numerous causes like water scarcity and lack of prey. Water scarcity is the primary cause followed by the frequent movement of tigers. In other ranges, very few cases occur due to the efficient monitoring system. In Kaila Devi, the bear population is quite higher resulting in bear–man conflicts in most of the ranges of Kaila Devi areas.

But one thing we notice is there were very fewer cases happened of human-wildlife conflicts due to some wild animals being very less rather than RNP, but the government is trying to set down critical tiger habitat (Fig. 14.7).

The reason for this is a lot of villages are situated in core areas of CTH, and some big towns are also located very near to forest boundary. Water scarcity is a large reason for the movement of wild animals. Naturally wild animals escape their core areas and conflict occurs (Figs. 14.8 and 14.9).

The highest number of cases of conflicts are with Leopard in comparison to other animal attacks in RNP. Cases by Tigers are on 2nd number. Other wild animals also proved more lethal in the RNP range. Few cases were reported for Bear. This reflects that in the RNP Area the concentration of Leopard and Tiger is higher than the other animals.

In the core area of RTR, there are so many cases registered against poaching activities, especially one of the very well-known traditional hunter communities "Mogiya" which is always involved in poaching, according to the data, most poaching

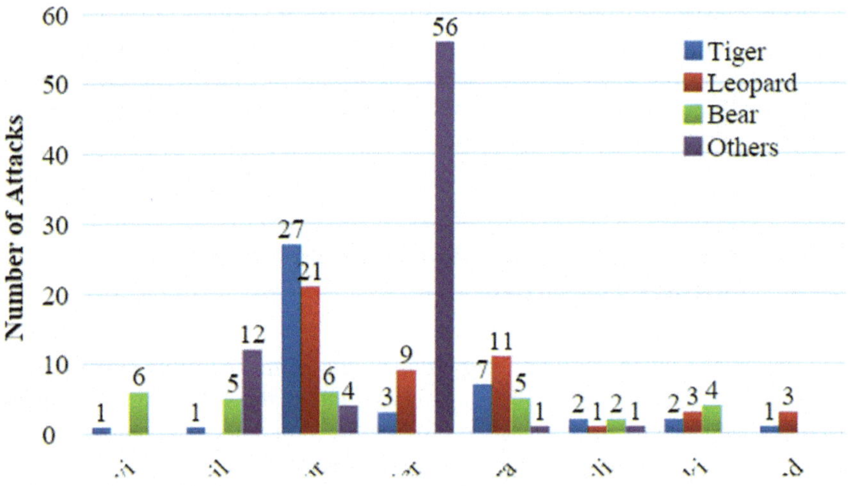

Fig. 14.7 Range-wise human-wildlife conflict in Karauli Division, 2001–2016 (*Source* Forest Department, RTR, 2016)

Fig. 14.8 Anthropogenic pressure in core and buffer areas of Ranthambhore Tiger Reserve (*Source* Compiled by scholars, 2016)

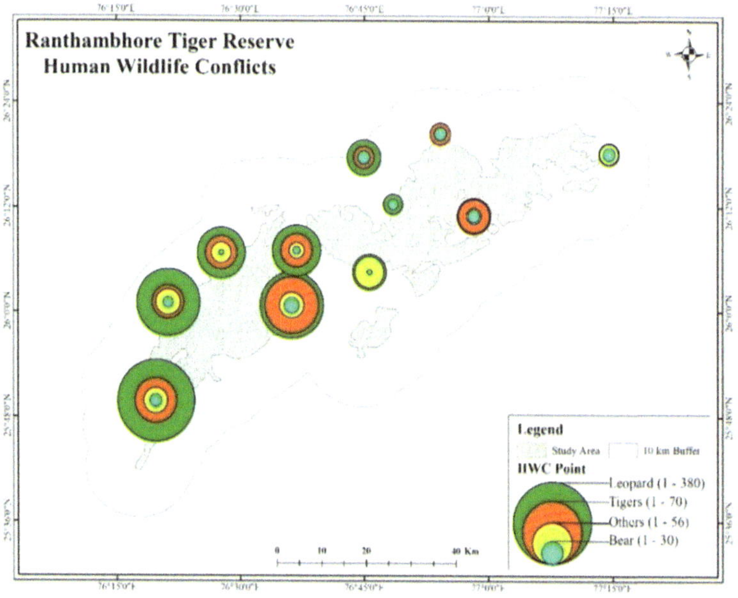

Fig. 14.9 Human-wildlife conflicts in RTR &10 K.M. Buffer areas (*Source* Forest Department, RTR, 2016)

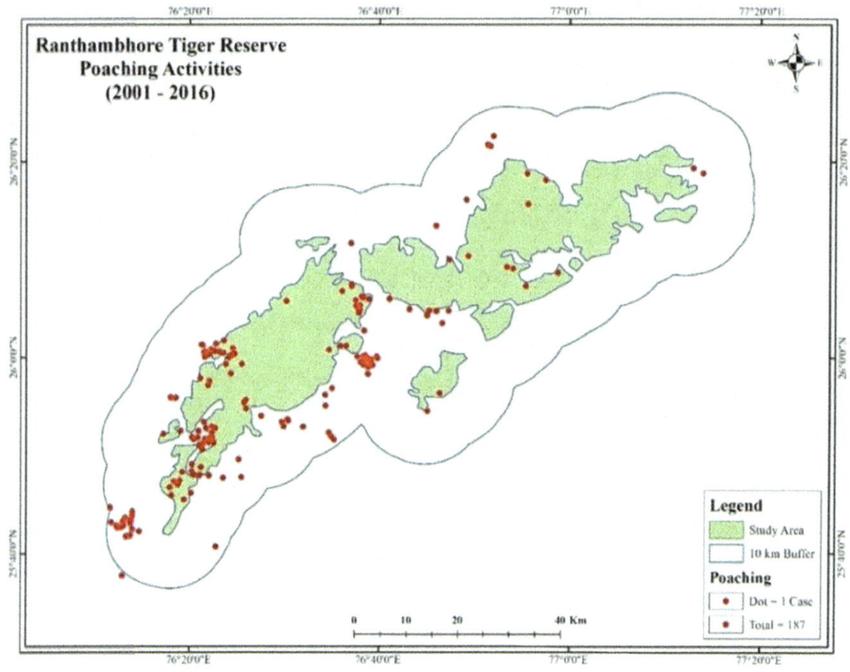

Fig. 14.10 Poaching activities in core and buffer areas of RTR 10 K.M. Buffer areas (*Source* Forest Department, RTR, 2016)

activities happened in RNP and BRHB, and Khandar range and few cases registered in KWLS also. Most cases are registered from the transition zone of the core and buffer region. Mostly poaching activities happen in the RNP range and few cases registered from Kela Devi Sanctuary (Fig. 14.10).

14.5 Conclusion

In the Ranthambhore Tiger Reserve from 2001 to 2016, there are increasing human-wildlife conflicts day by day due to numerous causes such as increasing tiger and leopard population, human encroachment, and land use land cover changes. During monitoring and movement of wild animals, water scarcity, men-eating habit, domestic cattle eater habits, lack of area for tiger territory, pilgrimage activities, used of pasture land fodder land, and other reasons also. Eventually man-made and natural causes are responsible for pushing the conflict rate.

The most important issues of Ranthambhore are a human-wildlife conflict and voluntary village relocation which are instigate by tiger movements, fragmentation of corridors and other factors (Singh et al. 2015). Other issues include intelligence-based enforcement, anti-poaching campaigns, voluntary relocation of villages, special monitoring of tigers, and regulation of tourism as per guidelines issued by the National Tiger Conservation Authority (NTCA), environmental development in buffer areas to list. Support of local public to reduce forest resource dependency, early payment of ex-gratia amount for loot due to wild animals, security measures in areas having roads and power transmission lines, besides coordinating with the state of Madhya Pradesh and forest divisions within the state as well as to keep an eye on moving tigers.

The result underlines the crucial discussions on human-wildlife conflict in the core and buffer areas. Firstly, the focus is on secondary data which was collected from the district office of Ranthambhore, Sawai Madhopur. It covers almost 16 years of data which began from 2001 to 2016. According to data, some cases are continuously increasing but in the last two years conflict rate is decreasing but again it rose in 2016. Therefore, the conflict rate is moving up toward an ecological crisis which is more dangerous for both existence of humans and wild animals (Fig. 14.10).

According to seasonal data, the summer season has the highest case registered after the monsoon and winter season in Ranthambhore national park because in summer there is a scarcity of food, fodder, and shelter, therefore, wild animals migrate. During the monsoon and the winter season, there is rarely any shortage. While in Kaila Devi different conditions were leading to the highest number of cases registered during winter and followed by monsoon and the summer season. Thus as the human habitat is much bigger than Ranthambhore National Park, human encroachment is rapidly increasing in the core and buffer areas as well as "Dang areas". So that's why human interference is higher than RNP and creates a conflict. Otherwise, the human-wildlife conflict rate is higher in summer after rainy and winter which is shown in the RNP conflict rate. Science-policy and community interface will ensure sustainable forest resource management and provide a synoptic view of the forest service system for the better analysis of the issues in this region.

Acknowledgements The author wanted to thank you to my mentor and colleagues, who helped me directly or indirectly to conduct this scientific research work. Author expresses my heart full indebtedness and owes a deep sense of gratitude to my all well-wishers for their encouragement and consistent support. He is extremely thankful to Late Prof. R. B. Singh, Prof. Anjan Sen, Dr. Manjit Singh, Dr. Ritika Prasad, and Mr. Ravi Mishra for their sincere guidance and assistance in completing this paper. The author also thank all my friends who have contributed to the preparation of this research paper.

References

Adger WN (2006) Vulnerability. Global Environ Change 16(3):268–281

Baraso S (2012) An analysis of landscape characteristics influencing livestock depredation by Lions, Hyenas, and Leopards in Loibor Siret, Tanzania. Doctoral dissertation, Duke University

Chandy S, Euler DL (1998, October) Can community forestry conserve tigers in India? In: Personal, societal and ecological values of the wilderness: Sixth World Wilderness Congress proceedings on research, management and allocation, vol 2, pp 155–161

Conover MR (2001) Resolving human-wildlife conflicts: the science of wildlife damage management. CRC Press
Dickman AJ, Hazzah L (2016) Money, myths and man-eaters: complexities of human-wildlife conflict. In: Problematic wildlife. Springer, Cham, pp 339–356
Everatt KT, Kokes R, Pereira CL (2019) Evidence of a further emerging threat to lion conservation; targeted poaching for body parts. Biodivers Conserv 28(14):4099–4114
Gilleland AH (2010) Human-wildlife conflict across urbanization gradients: spatial, social, and ecological factors
Graham K, Beckerman AP, Thirgood S (2005) Human–predator-prey conflicts: ecological correlates, prey losses and patterns of management. Biol Conserv 122(2):159–171
Gurung B, Smith JLD, McDougal C, Karki JB, Barlow A (2008) Factors associated with human-killing tigers in Chitwan National Park, Nepal. Biol Conserv 141(12):3069–3078
Inskip C, Zimmermann A (2009) Human-felid conflict: a review of patterns and priorities worldwide. Oryx 43(1):18–34
Messmer TA (2000) The emergence of human-wildlife conflict management: turning challenges into opportunities. Int Biodeterior Biodegrad 45(3–4):97–102
Mishra C (1997) Livestock depredation by large carnivores in the Indian trans-Himalaya: conflict perceptions and conservation prospects. Environ Conserv 24(4):338–343
Mishra C (2001) High altitude survival: conflicts between pastoralism and wildlife in the Trans-Himalaya
Mishra C, Allen P, McCarthy TOM, Madhusudan MD, Bayarjargal A, Prins HH (2003) The role of incentive programs in conserving the snowleopard. Conserv Biol 17(6):1512–1520
Msoroka MS, Amundsen D (2018) One size fits not quite all: universal research ethics with diversity. Research Ethics 14(3):1–17
Nasi R, Brown D, Wilkie D, Bennett E, Tutin C, Van Tol G, Christophersen T (2008) Conservation and use of wildlife-based resources: the bushmeat crisis. Secretariat of the Convention on Biological Diversity, Montreal and Center for International Forestry Research (CIFOR), Bogor. Technical Series, 50
Nyhus PJ, Tilson R (2004) Characterizing human-tiger conflict in Sumatra, Indonesia: implications for conservation. Oryx 38(1):68–74
Ogra M, Badola R (2008) Compensating human-wildlife conflict in protected area communities: ground-level perspectives from Uttarakhand, India. Human Ecology 36(5):717–729
Redpath SM, Young J, Evely A, Adams WM, Sutherland WJ, Whitehouse A, Gutierrez RJ (2013) Understanding and managing conservation conflicts. Trends Ecol Evol 28(2):100–109
Schwerdtner K, Gruber B (2007) A conceptual framework for damage compensation schemes. Biol Cons 134(3):354–360
Sillero-Zubiri C, Laurenson MK (2001) Interactions between carnivores and local communities: conflict or co-existence? Conserv Biol Ser-Cambridge:282–312
Sillero-Zubiri C, Sukumar R, Treves A (2007) Living with wildlife: the roots of conflict and the solutions. Key Top Conserv Bio:266–272
Singh R, Nigam P, Qureshi Q, Sankar K, Krausman PR, Goyal SP, Nicholoson KL (2015) Characterizing human-tiger conflict in and around Ranthambhore Tiger Reserve, Western India. Eur J Wildl Res 61(2):255–261
Thirgood S, Woodroffe R, Rabinowitz A (2005) The impact of human-wildlife conflict on human lives and livelihoods. Conserv Biol Ser-Cambridge 9:13
Treves A, Karanth KU (2003) Human-carnivore conflict and perspectives on carnivore management worldwide. Conserv Biol 17(6):1491–1499

Chapter 15
Environmental Ethics: In the Context of the COVID-19

Pran Kumar Rajak

Abstract There is an intimate relationship between humans and the environment. If people use the environment well then they will get good results and if used badly then environment will not give the same. It's not just that people have a moral relationship with people. The aspect that has a moral relationship with human beings as well as the whole living world and the environment is highlighted here and it has also been mentioned that the moral aspect of the environment with human beings is not to be neglected in the corona atmosphere but to adhere properly for the overall welfare. All the changes that have taken place in environmental ethics as a result of living in a long Corona period. (1) People have become more selfish. (2) The poor have become poorer. (3) Children are house prisoners, unable to enjoy childhood. (4) The women's entertainment of the neighborhood can no longer be seen. (5) The number of unemployed has increased. (6) Many people have lost their livelihood. (7) There is absolutely no hospitality in the society. (8) The situation is very bad as the street dogs do not get to eat.

Keywords Society · Rich-poor · Children-women · Livelihood · Non-anthropocentric-view

15.1 Introduction

There is an intimate relationship between the two—humans and the environment, and there is an interaction. If people use the environment well then people will get good results from environment and if people use environment badly then environment will not give good results to people. We, the people, will cut down the trees around the house and make it a playground, cut down the forest and build a city and as a result nature will not love us, it will cut off the supply of oxygen which

P. K. Rajak (✉)
Department of Philosophy, M. U. C. Women's College, Purba Bardhaman, West Bengal 713104, India
e-mail: prajak1974@gmail.com

A. Dubey et al. (eds.), *Promoting Sustainability Through Water Management and Climate Change Adaptation*, Advances in Geographical and Environmental Sciences,
https://doi.org/10.1007/978-981-99-5479-7_15

is very important for our survival. This is what is supposed to happen 'As action, so result'. In Indian philosophy, Vedic philosophy like Nyāya-Vaiśeṣika, Sāṁkhya-Yoga, Mīmāṁsā-Vedānta, etc., and non-Vedic philosophy like Jainism, Buddhism, etc., accept Karmavāda. Karmavāda says: every living being must enjoy the result of his deeds. We have always had a firm belief that no matter how badly we treat nature, nature will never treat us badly. We thought nature would meet all our needs, so we continued to use nature as we were. But nature will tolerate our actions forever and forgive us—it is impossible. Nature has started her game because it is impossible and as a result she has been teaching people by keeping them under house arrest for so long. Today we are bound by the Corona atmosphere.

15.2 Objective

It's not just that people have a moral relationship with people. The aspect that has a moral relationship with human beings as well as the whole living world and the environment is highlighted here and it has also been mentioned that the moral aspect of the environment with human beings is not to be neglected in the corona atmosphere but to adhere properly for the overall welfare.

15.3 Methodology

Some books have been used to create this article and in many cases have to rely on the surrounding conditions.

15.4 Discussion

Major Findings: Environment and Ethics—The totality of our earth's seasons, day-night rotation, waterfalls, rivers, mountains, forests, grasslands, animals, birds, light, air, warmth, etc., is the natural environment of the earth and every living and strange thing on earth. "Environment (biology definition): (1) the external conditions, resources, stimuli, etc., with which an organism interacts, (2) the external surroundings including all of the biotic and abiotic factors that surround and effect the survival and development of an organism or population, (3) the totality of the surrounding conditions and elements in an individual" (Johnson et al. 1997).

The philosophical discussion of morality is ethics. Ethics is the science that evaluates human character or behavior. Ethics discusses philosophically about human behavior, usage, customs, and principles. Mackenzie said, "*the term 'Moral Philosophy', which means the same thing as Ethics, is derived from the Latin mores, meaning habit or customs. Ethics, then, we may say, discusses men's habit and customs, or in other words their characters, the principles on which they habitually act, and considers what it is that constitutes the rightness or wrongness of those principles, the good or evil of those habits*" (Mackenzie 1929). In this case, words like 'justice', 'injustice', 'good', 'evil', 'appropriate', 'inappropriate', 'moral', 'immoral', etc., are used to evaluate human behavior. According to Lillie, "*We may define ethics as the normative science of conduct of human beings living in societies-a science which judges this conduct to be right or wrong, to be good or bad- or in some similar way*" (Lillie 1955). Ethics is judged on the basis of an ideal. This ideal is, according to Bentham and Mill, happiness and according to Kant, the moral law.

Ethics is not discussed separately in Indian philosophy, inherent in the discussion of morality, religion and philosophy. Ethics speaks of rules that help in self-purification and social welfare. People follow moral rules, for liberation. Liberation is the ultimate goal of the whole nature, which can be achieved only from complete selflessness, from selfless deeds. "*Liberation, according to Ramānuja, is not the merging of the individual soul into the Absolute, but only the direct intuitive realization by the individual soul of its own essential nature as a mode of God*" (Sharma 1960). "*The liberated man is the ideal of society and his life should be worthy of imitation by the people at large*" (Chatterjee and Dutta 1968).

Environmental Ethics—The concept of environmental ethics, however, is not a very ancient concept. It can be said that the concept of environmental ethics has been awakening in the minds of the people since the seventies of the twentieth century. "*An environmental ethic would find virtue in saving and recycling resources, and vice in extravagance and unnecessary consumption*" (Singer 2000). From such an idea, philosophers have come to think that the natural environmental concept of the environment in which man lives is related to the discussion of morality. We are now forced to realize the fact that we, as human beings, have some moral responsibility toward the natural environment. Not only is the concept of environmental ethics confined to the thinking of scientists, but it has also stirred the minds of philosophers, ethicists and others. The human civilization that we are now observing around us is all the result of advances in science and technology. However, although this civilization is recognized as the result of progress, it does not contain and carry progress. On the one hand, people are trying to establish this progress, on the other hand, they are arbitrarily destroying natural resources and disturbing the balance of nature. As a result, the natural environment of man is full of crises and the concept of environmental ethics has been created for the purpose of getting rid of this crisis.

More than One and a half years have passed and we are constantly moving toward an unknown future through panic. We are now spending the day in a closed environment for the Corona situation. The effect of Corona is not decreasing at the rate at which it is increasing upwards. Now it is heard that this time Corona can have a huge impact on children. This is even more alarming because the Corona vaccine has been introduced in adults but the Corona vaccine for children has not yet been discovered. All the changes that have taken place in environmental ethics as a result of living in a long Corona period. (1) There is absolutely no hospitality in the society. (2) Person-to-person conversations are no longer visible, and social and emotional distance has increased. (3) People have become more and more selfish. (4) The poor have become poorer and the rich have become richer. (5) Many people have lost their livelihood. (6) The number of unemployed has increased. (7) Children are house prisoners, unable to enjoy childhood. (8) School dropouts have increased in the field of education. (9) The women's entertainment of the neighborhood can no longer be seen. (10) The boys no longer hang out at neighborhood clubs. (11) The situation is very bad as the street dogs do not get to eat. (12) Medicinal plants are being severely damaged during this time. (13) The communication system is irregular.

Humans are social creatures. Living in a society, a person forms some social groups, institutions and relationships with the person through which the daily activities and behavior of the person are conducted. That is, they work together on the basis of mutual relations and cooperation to achieve any goal. Society is essential to human beings. People cannot live alone without the company of others. But the current situation in Corona is just the opposite. Person-to-person conversations are no longer visible, and social and emotional distance has increased. MacIver and Page said, "*Society is a system of usages and procedures, of authority and mutual aid, of many groupings and divisions, of controls of human behaviour and of liberties*" (MacIver and Page 1974). Social relationships can be of different types, namely family, personal, moral, logical, impersonal, economic, political and others.

Moral Obligations (society, non-society)—Indian philosophy is deeply spiritual and a philosophy of truth or theory. The main point—renunciation and asceticism is the goal of human life, the ultimate salvation. People live morally with the attitude of sacrifice. Every human being is born into a society with certain debts and moral obligations. If man does not repay those debts, then the world powers will not help him in his liberation. Whatever happens in the field of environmental ethics, we do not help the environment in return for what nature gives us so much. Rather we, humans oppress nature and we oppress various living creatures and plants. Let that natural environment be normal and let the corona be affected. Over a long period of time, street dogs have gradually fallen ill and died from malnutrition, which has affected or will affect the environment. What people think about this, if they do, is very insignificant. We are indebted to nature. The word 'debt' means 'liability'. Every human being in society is dependent on others in one way or another, those other people can be of past and present, there may be other organisms and inert matter. People are indebted to everyone in many ways and morally and socially people are responsible to repay the debt of all.

Five Sacrifice—In order to repay the debt (ṛṅa), people make some offerings to the lender through the PañcaMahāyajña (five sacrifice)—sage sacrifice, patriarchal sacrifice, devotional sacrifice, human sacrifice and ghost sacrifice (ṛṣiyajña, pitṛyajña, dev yajña, nṛ-yajña and bhūtayajña). The householder, the sage performs by studying and teaching the Vedas, the patriarch performs by donating water for the purpose of the ancestors, the deity performs by donating invocation to the fire. The process of repaying debts by taking care of all kinds of animals and plants is ghost sacrifice.

Human Sacrifice (livelihood, rich-poor, children-women)—The householder performs nṛ-yajña or manuṣyayajña through hospitality. There are ways in which guests can be entertained—by donating food to the hungry, by donating clothes to the naked, by donating houses to the homeless, by repaying debts through services to the sick. All these ways have been heard in Indian society since ancient times. But now, in the Corona situation, these ways do not exist in the view of antiquity. Some voluntary organizations have distributed cooked food in slum areas in Corona situations, sometimes distributing old clothes. Some organizations try to put up photos as evidence. People who have the ability to entertain guests personally are also trying to get richer and richer by being selfish at this time without thinking about entertaining guests and the poor are losing their livelihood and becoming poorer, and the number of unemployed is increasing. People have become more and more selfish in Corona situations. Socially, physically and mentally the distance between the person and the person is increasing. Most of the time when a person sees an acquaintance around the house, he pretends not to see.

For more than one year and a half, children have been trapped inside the house in Corona. Unable to enjoy their childhood for so long, they inadvertently irritate their mother. If the financial condition of the house is good, the mother often calms the child with toys and food items. But how will the mother calm down the child whose father is the gateman of the bus, hawks on the train, runs the canteen in any school or college? Many people have lost their livelihood due to the closure of buses and trains due to long days of lockdown and severe loss of transportation, many people have lost their livelihood due to the closure of schools and colleges. Many women used to cook in other people's houses, but due to Corona's situation, they stopped working. Because of Corona, the lives of these poor people have changed, the poor people have become even poorer. The mother cannot give her children normal food; nutritious food is the next thing. The mental state of poor mothers is so bad that they do not have the strength to sit, talk and discuss even while maintaining equal distance with their neighbors. How can poor children, who do not have a minimum of food at home, join classes on their smart phones? That is why they are not able to attend online classes. Because of the Corona situation, friends are not able to hang out together or play sports, tell stories or discuss in the club. Poor people do not have the ability to see a doctor or buy medicine with money if they are sick, they can be cured if they can get medicine from the hospital free of cost. Many people drink it like tea with basil leaves, bask leaves and ginger in boiling water. All these medicinal plants are damaged as a result of removing too many basil leaves and bask leaves.

15.5 Conclusion

People usually cannot fully benefit. We see in daily life, many virtuous people are unhappy, he have no happiness in the life of the family, he has no joy/pleasure but he has done his duty for the rest of his life. Although this duty is completely man's own, happiness or joy/ pleasure is not in our own hands. We cannot be happy as we wish, even if we do our duty with devotion. Happiness or joy/pleasure in most cases depends on the external environment while the duties of the person depend on the person. If the external environment is right, we, the people will be better, we will be able to enjoy happiness and joy. There are other animals and plants besides humans in the external environment, if we can faithfully fulfill the moral responsibilities that we have toward them, the environmental ethics will be maintained, nature will be healthy and normal, there will be overall welfare. In the current Corona situation, we should not be unduly alarmed, but we should be able to maintain a courtesy of moral dignity toward human beings, other animals, plants and the entire material world in accordance with the governmental/official health rules.

References

Chatterjee S, Dutta D (1968) An introduction to Indian philosophy. University of Calcutta
Johnson DL, Ambrose SH, Bassett TJ, Bowen ML, Crummey DE, Isaacson JS, Winter-Nelson AE (1997) Meanings of environmental terms. J Environ Qual 26(3):581–589
Lillie W (1955) An introduction to ethics. Methuen & Co Ltd.
MacIver RM, Page CH (1974) Society: An introductory analysis. Macmillan India Limited
Mackenzie JS (1929) A manual of ethics. University Tutorial Press Ltd.
Sharma C (1960) A critical survey of Indian philosophy. MotilalBanarsidass Publishers Private Limited
Singer P (2000) Practical ethics. Cambridge University Press

Chapter 16
Building Sustainable Livelihoods Through Everyday 'Green Urbanism' Practices—A Case Study of GTB Nagar Neighborhood Delhi, India

Anisha, Komal, Sneha Agrawal, and Vaishali Tomar

Abstract As urban processes become increasingly complex over time, cities expanded at the cost of green spaces. "The blind forces of urbanization flowing along the lines of least resistance", as argued by Mumford (1956) "show no aptitude for creating an urban and industrial pattern that will be stable and self-sustaining". Through the age of "The Anthropocene", cities evolved from their traditional socio-political-economic functions to geological actors with the potential capacity to alter the earth's climate- locally and globally. This has made urban a more difficult place to live initiating an urge to look for more innovative ways to improve the quality of city life by bringing nature back into everyday living spaces. People began greening the congested urban spaces to make a healthy living in the cities possible. This "green turn" in urbanization has taken various forms across the world. Indian cities are influenced by green turns in urbanization. In this context, the paper tries to map and comprehend home gardening practices as everyday negotiation with the urban crisis in Delhi. The study grounded on both primary and secondary data, and involving a mixed methodology tries to focus on the untold but practiced epistemologies of different forms of green urbanism in Delhi.

Keywords Urban crisis · Green urbanism · Urban gardening · Sustainability · Green turn

Anisha (✉)
Department of Geography, Delhi School of Economics, University of Delhi, Delhi 110007, India
e-mail: anishabo09@gmail.com

Komal · S. Agrawal · V. Tomar
Indraprastha College for Women, University of Delhi, Delhi 110054, India

A. Dubey et al. (eds.), *Promoting Sustainability Through Water Management and Climate Change Adaptation*, Advances in Geographical and Environmental Sciences,
https://doi.org/10.1007/978-981-99-5479-7_16

16.1 Introduction

Historically "cities are the man's most successful attempt to remake the world he lives in more after his heart's desire", but over time with congestion, pollution, inequality, social segregation and gentrification, etc., they have emerged as a challenging place to live (Park 1967 as quoted in Harvey 2008). The discourses on sustainable and resilient cities emerged as a response to this scenario. The Green Urbanism movement was one of the prominent alternative discourses. It emphasizes the important role of cities and positive urbanism in shaping more sustainable places, communities and lifestyles (Beatley 2000).

In the contemporary Asian Urban Century, as a major part of the urban stories are unfolding in the global south, the urban problems are even more complex and multi-faceted. Home gardening and everyday green urbanism have emerged as prominent sites of interest in urbanism and policy discourses in developing countries including India. They are confronted with the question of unprecedented demographic change and infrastructural planning. As Roy put it, "Third World Megacities defy all norms of rational planning and future-proofing" (Roy 2009).

In this context, the paper examines how the different constituents of Indian urbanism negotiate with the environmental constraints of their congested cityscapes. The paper, by engaging with people's sustainability practices argues that the varied forms of negotiation the actors make within the confines of their cultural milieus affect the way they shape the life of the city, and the associated production of nature in those contested urbanscopes. The paper also looks at the role of local geographies in this process. The findings of the article rely primarily on fieldwork material. The research applied the mixed method of analysis, by bringing qualitative and quantitative methods together to make better sense of the issues.

16.2 Data and Methods

The data used in the study are both primary and secondary in nature. Primary data includes field surveys and various secondary sources of data are obtained from a government report, DDA master plan, Census of India, and newspaper articles.

The methodology is mixed in nature. Both qualitative and quantitative methods had been used. The techniques of content analysis and discourse analysis have been thoroughly used to get a deep insight into the matter.

16.3 Methodological Framework

16.3.1 Study Area

The study was conducted in GTB Nagar Neighborhood, Delhi one of the fastest-growing areas in terms of urbanization. The area has witnessed an increase in home gardening practices even in the congested apartment in the last few years in spite of the fact that it is well-planned in terms of open spaces and community parks. The local geographies of this area are one of the main reason which initiated an urge to bring nature back into their everyday life (Fig. 16.1).

16.4 Results, Discussions, and Analysis

Gardens have always been places of ambivalent sensibilities creating links between natural and domesticated environments. Urban dwellers in the desire of beautifying their homes and as a coping mechanism to urban crisis adopt urban gardening practices which create a positive influence on the urban environment. It provides innovative solutions to enhance the quality of urban space, encourages sustainable lifestyles, and positively impacts the well-being of urban dwellers. However, the spaces and

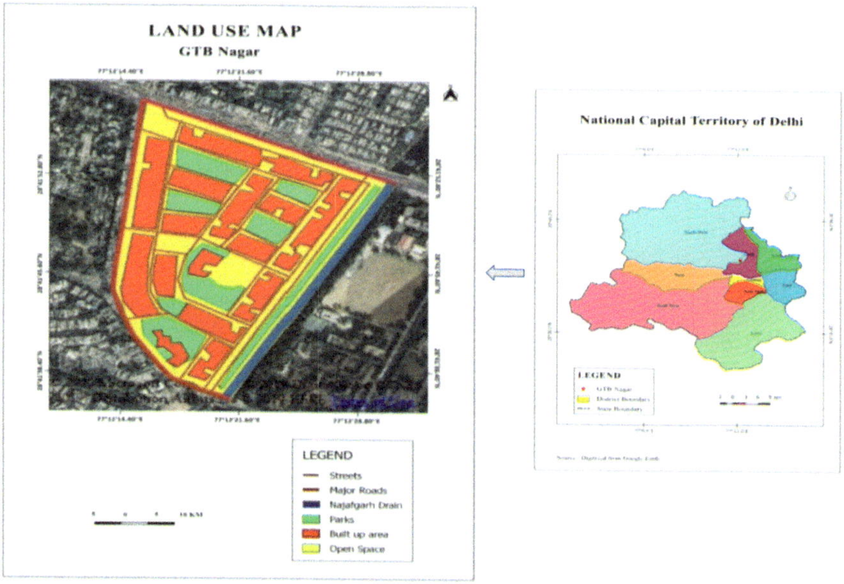

Fig. 16.1 Land use map of study area

their meanings vary widely according to the socioeconomic and political context in which they are based.

16.4.1 Organization of Home Space

The organization of space for home gardening practices in the congested apartment of Delhi by urban dwellers is the consequence of the urban restructuring processes, evolving social conditions, and inaccessible public green space. The study shows that the gardening practice follows the distance decay mechanism phenomena associated with Najafgarh Drain. It has been observed that as the distance of the households from the Najafgarh drain increases the people's negotiation with nature decreases which suggests that people living near the drain are more engaged in gardening as compared to the people living far from the drain. This indicates the distinct spatial niches of green urban practices in cities like Delhi, which are conditioned not only by the urban restructuring process but also by local geographies such as the sense of geographies of smell and immediate pollution around the Najafgarh drain (Fig. 16.2).

Fig. 16.2 Organization of space in study area (*Source* Field Survey, 2019–2020)

Fig. 16.3 Methods of gardening (*Source* Field Survey, 2019–2020)

The notion of spatial pattern determines the choice of places for gardening. This choice of place is negotiated with the local conditions of the area. In this locality people mainly use the courtyard, balcony, and rooftop for gardening because these are the only available places in their apartments. As stated by Hackenbroch, "in light of new urban configurations and the changing social dynamics of urban space, conceptual and analytic attention has been given to the access to space as a product of negotiations".

Availability of place determines the methods used for gardening. In the study area, people are using containers (65%), ground (10%), and raised bedding (25%) for the plantation. Some people are using hanging pots and some other methods as well to keep the pots which show that people are creating places to stay in touch with nature. This indicates that spaces and their meanings vary widely according to the socioeconomic and physical context in which they are based (Fig. 16.3).

16.4.2 Actors, Meanings, and Perceptions

The meaning and perceptions of gardening vary among the different categories of urban dwellers based on the varied socioeconomic conditions as well as on their forces of motivation. From our survey, it has been found that members of the house are active participants but around 38.7% of people are taking the help of a gardener. Gardeners, vendors, and home maids are also prominent actors in keeping this practice alive in the urbanized world. One gardener pointed out, "The scenic beauty of the city depends on poor working hands". In the process of beautifying the homes, they constitute an important agent of urban environmental sustainability. Gardeners are usually employed for performing those tasks which are risky and harsh like the application of fertilizers, fencing, and plowing. Activities like site and seed selection, watering, and raising, are performed by the family members (Fig. 16.4).

Over the past few years, research on the gendered nature of gardening has demonstrated the existence of different attitudes of men and women toward gardening

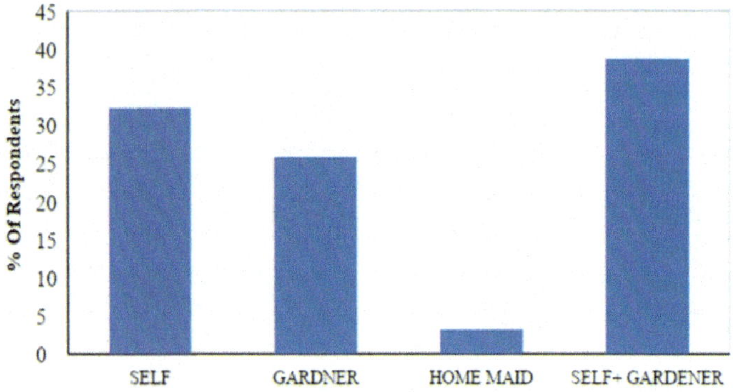

Fig. 16.4 Garden management (*Source* Field Survey, 2019–2020)

behavior. The present study demonstrates the gendered nature of gardening where women are more involved. This is more evident in the following two contrasting narratives from the field:

> I don't have any idea about their gardening. I can only tell you how much I pay to the gardener. (An employed male respondent, 35 years old)

> I find it as a form of connecting me to nature and it gives me a kind of spiritual experience. That is the most relaxing period when I spend with my plants. (A retired female professor)

This could be used to testify from the eco-feminists argument that there is a connection between women and nature that emerged from their shared history of oppression by patriarchy and western society (Shiva and Mies 1993). Accordingly, women are environmentalists by default and play a key role in bringing ecological balance. However, the field survey does not show that men are absent from gardening practices. Rather there are distinctly gendered roles men and women play in gardening practices. Men, either the family member or the gardener, are engaged in activities such as land selection, preparation, trellising, fencing, and application of fertilizers and pesticides. Female family members are mainly involved in beautifying the garden spaces and their everyday maintenance, which are more tedious tasks requiring high levels of patience. The present study also found that non-working women are more involved in this practice. The social role of women as a main unpaid domestic worker in each household brings them closer to an awareness of environmental hazard. Thus, we can say that to some extent women are expected to take up gardening as a result of gendered divisions of labor within the household (Fig. 16.5).

It is well evident from the survey that the meaning of gardening changes with age groups of urban dwellers. People above the age group 40 are predominantly involved in gardening. The most probable reason is that they are the ones more vulnerable to the pressure of the urban crisis and thus they look at gardening as an alternative to attain mental peace and better health and living. Moreover, people in the 60-plus age group are in retirement age, and therefore gardening constitutes an important means to spend their time. This is termed as retirement gardening when people choose gardening as their hobby after retirement.

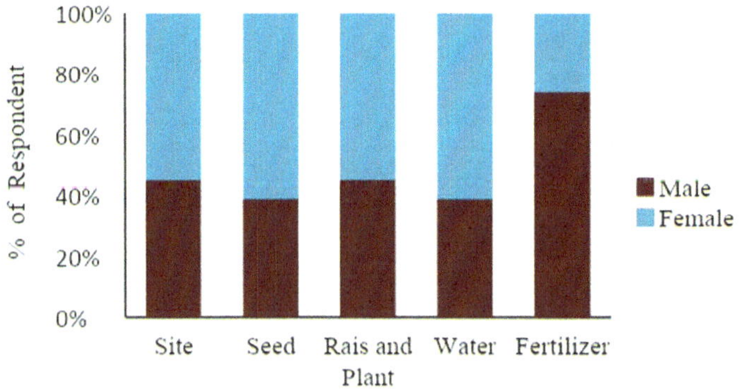

Fig. 16.5 Gender involvement in gardening (*Source* Field Survey, 2019–2020)

Young generations are also involved to some extent in watering and planting due to their growing awareness of the environment in recent years and also because of their family instinct toward these practices. One student said, "I do gardening because I was taught from childhood that it is important to get in touch with nature. It all depends on upbringing". But most of the younger generations still don't understand the value of such little steps. They considered it time-consuming, and worthless.

16.4.3 Motivations Behind Gardening

Variations in the pattern of home gardening practices are the product of the motivation of urban dwellers. Field observations suggest that the major motivational factors involved in gardening practices are: aesthetic needs, psychological needs, or for food requirements. In GTB Nagar, the local geographies, mainly Najafgarh drain play an important role in determining the people's choice of the plant variety. Plants' preferences along with the increasing distance from Najafgarh drain are shown in the form of scatter plots in Fig. 16.6.

16.4.4 Gardening and Sustainability

Sustainable gardening is primarily aimed at growing plants in such a way that keeps the soil, and plants alive and healthy by using proper and efficient resources, i.e., water, organic fertilizers, and household wastes efficiently. It is thus a tool for sustainable development as it addresses a wide array of environmental, economic, and social objectives. In the GTB Nagar locality, most of the respondents are using organic seeds and fertilizers. However, we cannot ignore the fact that access to sustainable products

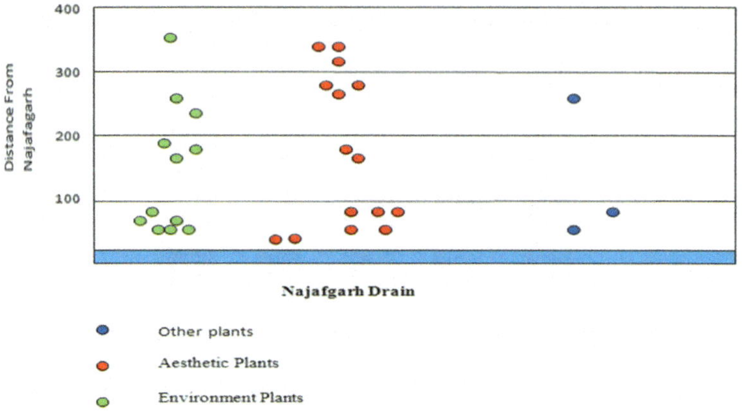

Fig. 16.6 Transect chart showing plants preferences vis-à-vis distance (*Source* Field Survey, 2019–2020)

is related to the economic status of the people and advice from the gardener, and the levels of awareness among urban dwellers (Fig. 16.7).

Another important aspect of sustainable gardening is the efficiency of water use. It has been observed that most of the respondents are using tap water as the source for watering the plants. There is a lack of awareness regarding the efficient method of irrigation. The usage of pots for gardening is an essential necessity in urban gardening due to the lack of space in GTB Nagar. But the trend of plastic pots and cemented pots over earthen pots directs such practices toward unsustainability. The prime reason for such change is that plastic pots and cemented pots are easy to maintain and it also enhances the beauty of the apartment. However indirectly the urban dwellers

Fig. 16.7 Fertilizer used in gardening (*Source* Field Survey, 2019–2020)

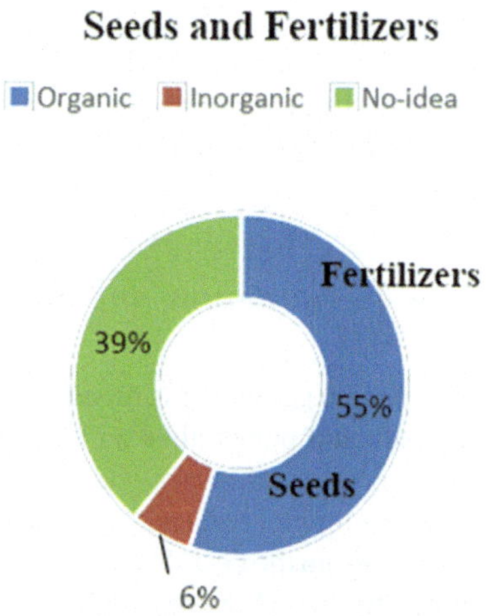

are burdening the city with heaps of plastics and causing stunted growth of plants as these pots are hard and less porous.

16.5 Conclusion

The study shows that even though the urban is a governmental category (Roy 2009), people through their everyday gardening practices try to contest the urban–rural binaries inherent in the governmental category. The attempts by people to bring nature back to the city spaces are not a homogeneous process. It has been conditioned by the local geographies. This has resulted in the making of vivid geographies of green urbanism in the locality. The social contexts of the actors involved add further complexities to the created geographies. The study shows that a large share of people are using organic fertilizers and seeds thus contributing to urban sustainability but they lack care while using water for irrigating the garden. This shows that people are open to "buying sustainability" if it's available from the market but find it harder to practice in their everyday life. This makes the question of sustainability at stake.

Therefore, we argue that green urbanism in the form of gardening practices is an attempt on the part of urban dwellers to negotiate the urban crisis in their everyday life. There are geographies of green urbanism shaped by the local context. Moreover, there is a need to bring the question of sustainability to the center of these practices. We believe, there is a need for the government to make proper policy interventions to further boost these individual-oriented and community-centric green practices to convert the "Asian Urban Century" into a century of sustainable urbanism.

Acknowledgements Authors are highly indebted to the Institute of Public Enterprise, OU Campus, Hyderabad for providing us the opportunities to publish our paper with Springer Nature Series. We express our sincere gratitude to Indraprastha College for Women, University of Delhi for giving us the suitable environment for the initiation of this research paper. Our sincere thanks to Balakrishnan P. Sir, Assistant Professor, Department of Geography, Indraprastha College for Women, University of Delhi for his exemplary guidance and valuable feedback throughout the course of this research. Also, Authors would like to thank Dr. Shrawan Acharya, Professor, JNU for giving his valuable time to us for interaction on our concerned work. His deep and valuable insights immensely helped us in organizing our work.

Finally, Author highly appreciate the kind nature of GTB Nagar residents who have helped us in successfully conducting our field-based survey for this research work.

References

Beatley T (2000) Green urbanism: learning from European cities. Island Press, Washington, DC

Harvey D (2008) The right to city. Int J Urban Reg Res 27(4):939–941

Roy A (2009) Why India cannot plan its cities: informality, insurgence and the idiom of urbanization. Plan Theory 8(1):76–87

Shiva V, Mies M (1993) Ecofeminism. Sage Publications Inc., London

Chapter 17
Understanding the Impact of Climate Change and Policy Development in India During Post-NAPCC Era

Nisikant Nayak and Sumanta Nayek

Abstract Climate change is a geo-natural and anthropogenic continuous process. It is regarded as one of the major concerning problems that the humanity faces today. Its gravity is more serious than any other global risks. However, the pace of its progress can be halted, and its impacts can be better adapted. In this context, two important strategic policy alternatives, 'adaptation' and 'mitigation' can be taken into account. Considering climate severity and all-encompassing implications, it warrants urgent collective policy formulation and responses. The primary objective of the study is to understand and identify policy gaps between policy formulation and its implementation and to suggest possible strategic policy recommendations to bridge the policy gaps to address the issue of climate change. The study primarily analyzes climate change policy issues and development in India since the initiation of the first ever National Action Plan on Climate Change (NAPCC 2008) policy of India. Climate change as an issue requires joint implementation of effective policies and programmes both by the Centre and its units. This study includes integrative methodological approach for climate policy analysis and implementation purposes. Furthermore, the study would compare the integrated policy at the national level (top-down) with region-specific (bottom-up) approaches while dealing with policy development, implementation and its effectiveness.

Keywords Climate change · NAPCC · Policy recommendation · Vulnerability analysis

N. Nayak (✉)
Department of Political Studies, Unitedworld School of Liberal Arts and Mass Communication, Karnavati University, Gandhinagar, India
e-mail: nisikant11@gmail.com

School of Social Sciences and Languages, Vellore Institute of Technology, Chennai, India

S. Nayek
Amity Institute of Environmental Sciences, Amity University, Kolkata, India

A. Dubey et al. (eds.), *Promoting Sustainability Through Water Management and Climate Change Adaptation*, Advances in Geographical and Environmental Sciences,
https://doi.org/10.1007/978-981-99-5479-7_17

17.1 Introduction

Climate change is now genuinely international policy concerns. Almost all countries are developing mechanisms to initiate domestic actions in sync with international policy regimes evolved through global climatic governance system (Methmann et al. 2013; Stevenson and Dryzek 2014; Held et al. 2014). The "Paris Climate Accord of 2015 signed under the aegis of the UNFCCC by 196 countries" is determined to keep "the global average temperature to well below 2 °C above pre-industrial levels." In this context it is relevant to ponder over the extent of global carbon emissions from fossil fuels (1900–2014); the global GHGs formations; the various economic sectors contributing the global emissions, and the shares of various countries contributing to global emissions, etc. Industrial revolution has caused to unprecedented changes in the atmosphere. As per the IPCC's third assessment report (2013), "human activities have increased atmospheric concentrations of GHGs and aerosols since the preindustrial era."

Carbon Dioxide (CO_2), the primary greenhouse gas (GHG), has increased to significant level from a pre-industrial level. Methane (CH4) concentration has gone up from 700 to 1,750 ppm. The Nitrous Oxide (N_2O) has increased from 270 to 316 during the same period. Similar is the case of troposphere ozone, which has increased by an average of 35% over that time. Its variation can be region-specific. Aside from the temperature records, the evidences of warming can also be observed in the case of increasing sea level (by 1–2 mm per year), shorter ice cover of river and lake (by two weeks), thinning of arctic sea ice (by 40% in thickness and 10–15% in extent), widespread retreat of non-polar glaciers, snow cover and permafrost (by 10%), longer growing seasons (by 1–4 days), pole-ward and upward shifts in plant and animal ranges, as well as earlier plant flowering, bird arrival breeding seasons, and emergence of insects and increasingly frequent coral reef bleaching (IPCC 2014; Dash 2007). In continuation to the above, it is also established that increased level of temperatures would trigger abrupt and large-scale changes caused by heavy rainfall, cyclone, severe flooding, droughts, agriculture failure, displacement, disasters, etc. The impacts are not evenly distributed, the poorest countries and the poor and marginalized will be stricken than the rich and developed world because of inadequate adaptive capacities and access to resources.

17.2 India and Climate Change

Climate change poses a very severe threat to the nation-building in developing and under-developed countries. India's national circumstances make it highly vulnerable to climate change. Due to large and diverse topography, climate and biosphere, demography and extended coastline, agrarian society, high poverty ratio, low-human development indicators, the country's vulnerabilities to climate change are significant. With the geographical area of 3.28 million Km^2, (2.3% of the global land mass, having 18% of the world's demography), it is now bracketed as one of the most climate-risk country in the world (NATCOM 2012). As per the country's first climate change policy document, NAPCC (2008), "it may alter the distribution and quality of India's natural resources and adversely affect the livelihood of its people." As per the *Global Climate Risk Index* (2017), India ranks fourth among ten major affected countries. The country witnessed huge unseasonal rainfall and suffered several deadliest heat waves in the world history. As a result, nearly 2,300 people succumbed to climate-induced disasters. Moreover, the modest incomes and lack of necessary facilities make a dent in the adaptation capacities of the society, hence forcing compromises to adapt to varied impacts of climate change. It is a considerable challenge to domestic public policy and governance capabilities.

By recognizing the urgency, the Union government has initiated a collective action to meet the objective of the Convention (NAPCC 2008; TERI 2009). By being part of several international agreements, the country has drawn the following points at the international levels. The country's role has been "(1) to become an influential voice and defender of the global south; (2) a producer of ideas and international norms and rules on climate change; (3) an effective coalition-builder; and (4) an aggressive protector of its own interests." The above roles reinforce each other (Agarwal and Narain 1991). "India is committed to engage actively in multilateral negotiations under the UNFCCC in a positive, creative and forward-looking manner. Our objective is to establish an effective, cooperative and equitable global architecture based on climate justice and the principles of Equity and Common but Differentiated Responsibilities and Respective Capabilities, under the UNFCCC. As we put together the new global compact for enhanced actions, it is critical to ensure that it is comprehensive, balanced, equitable, and pragmatic. It should address all the elements including Adaptation, Mitigation, Finance, Technology Transfer, Capacity Building and Transparency of Action and Support. At the same time, the absolute requirements of developing countries like India for an equitable carbon and development space to achieve sustainable development and eradication of poverty needs to be safeguarded."

"Adaptation to climate change is a long-term and continuing process" (NAPCC 2008). For a successful adaptation, "there is need of increasing the country's adaptive capacity both in the short term and the long run, by including measures such as economic diversification to reduce a country's dependence on climate-sensitive economic sectors" (NAPCC 2008; UNEP Guide Book 2011). The National Mission on Sustainable Habitat "seeks to promote sustainable habitats by increasing forest cover, improved energy efficiency of buildings, urban planning, management of solid and liquid waste, the modal shift toward public transport, and conservation through appropriate changes in the legal and regulatory framework".

Toward mitigating climate change, India has taken significant departures from the past. "It has now pledge voluntarily to UNFCCC for a 20–25% reduction in GHG emissions intensity (excluding emissions from agriculture) of Indian GDP by 2020 from 2005 levels" (BUR 2015). India's emissions would have risen to significantly higher levels in the absence of actions that fulfill the voluntary pledge of reducing emission intensity of GDP. The Government of India has increased the overall renewable energy capacity target by about five times from 35,776.96 MW (as of 31 March 2015) to 175,000 MW by 2022 (BUR 2015). To ensure fuel efficiency, environmental friendliness and cost-effectiveness, the country is promoting for Coastal Shipping and Inland Water Transport. The Union Government has announced the implementation of Jal Marg Vikas to enhance the inland waterways transport. Sagarmala Project is another initiative with the objective to augment port-led development and promote efficient transportation of goods (India's First BUR Report to UNFCCC 2015).

17.3 National Climate Change Policy and Effectiveness

India's climate policy can be best described as a complex exercise in the context of intents for rapid industrialization and expansive energy consumption along with persistent and widespread poverty across the country. Till 2000, its climate policy had more foreign policy orientations than addressing domestic concerns. However, this began to change from mid-2000 as the size of policymakers began to grow from traditional policymaking to more broad-based policymaking. From the gamut of the nodal Ministry of Environment and Forest, Ministry of External Affairs and Ministry of New and Renewable Energy, the policymakers also included specialized research agency, scientific administrators, and experts who underlined the importance of addressing climate change domestically. The results were the formulation of NAPCC (2008), the first ever climate policy document in the country.

NAPCC intends to address the issue more elaborately and systematically. Centre plays significant roles in mitigating GHGs as most of the energy-centric ministries are at its disposal. The eight missions as underlined in the NAPCC require sectoral interventions at the sub-national level. Under the federal frameworks, the Centre is constitutionally mandated to transfer funds from its disposal to states to meet horizontal and vertical imbalances. Such transfers can be made through FCs and Plan models. Resources can be directed to the states (federating units) through Central Assistance to State Plans, Centrally Sponsored Schemes and Central Plan schemes. Even though Centre played a dominant role in influencing and orchestrating the formulations of SAPCCs, in spite of this, the States have attempted to highlight their "local needs, locale environment and economics concerns" in their respective plans. For instances, "Odisha put the focus on the reducing the transmission and distribution losses; Sikkim puts focus on the water conservation from the Himalayas, Himachal Pradesh advocated for the payment of using ecosystem services" (Jogesh and Dubash 2015). It is observed from the analysis of SAPCCs that states are keen to put the local issues into the national map and advocated for how to address the environmental concerns at the state level through developmental planning with local climatic solutions. The capacity of states in putting an effective SAPCC emerged as significant concerns. "While many States emphasized on a locally driven plan, it appears that their nodal administrative departments were not well equipped to handle the plan formulation, hence multilateral and bilateral agencies were brought in to bridge the capacity constraints."

17.3.1 Role of Government and Other Stakeholders

The Twelfth Five Year Plan (2012–2017) has "laid down a detailed agenda for sustainable development. The plan notably emphasizes on intervention in governance, planning and various technologies" (India BUR Report to UNFCCC 2015) to initiate certain actions to implement the low carbon growth and development strategies in the country as per *Kirit Parikh Committee Report on Low Carbon Strategies for Inclusive Growth* (2011). LED bulbs and tube lights are the results of this plan. Further, the 12th Five Year Plan (2012–2017) "proposes setting up a National Wind Energy Mission (NWM), similar to the mission for solar energy. This mission includes various policies such as encouraging mixed land use for wind generation and agriculture, approaches for independent regulator and mechanisms for using the National Clean Energy Fund (NCEF) to finance development of local grids by state utilities and distribution companies" (India BUR Report to UNFCCC 2015).

17.3.2 Institutional Framework

Climate policymaking in India is a complicated process. It is involved primarily due to the country has severe climate identities. At first, India due to its geographical variability is more vulnerable to climate change. Being the second most populated country in the world, India is one of the major emitters of GHGs. Secondly, from the perspective of per capita emission level; India emits a low amount than the major countries of the world.

The NAPCC was established in 2008 with a broad objective of eight missions (NAPCC 2008). The mission includes national solar power development, energy efficiency, sustainable habitat, water mission, Himalayan ecosystem, sustainable agriculture, and strategic knowledge on climate change (NAPCC 2008). As a reaction to the formulation of NAPCC, India pledged in 2009 to reduce the emission intensity by 20–25% from the 2005 level by 2020 (Dubash 2016). The NAPCC was comparatively more active during the initial years of the formulation of this plan. It is evident from this fact that, this policymaking body has met eight times between August 2009 and February and did not meet until November 2014 (PIB 2013; Dubash 2016). The Indian Network on Climate Change Assessment (INCCA) has provided an assessment report on the impacts of climate change in the 2030s. The assessment report has taken into consideration four crucial sectors of the Indian economy such as agriculture, water resources, biodiversity, and health. Four climate-sensitive regions of India have been taken as a part of the analysis than the coastal areas and north-east region (INCCA 2010; Dubash 2016).

17.3.3 Policy Gaps, Suggestion, and Policy Recommendation

There are several gaps in climate policy and institutional frameworks of the country. The NAPCC has not given any GHG reductions target and its timeline. It was speculative. How it can be met and how the mission's approach will be implemented the NAPCC was utterly silent on those details. The extent of budgetary outlay will be allocated through which implementing ministries and how it can be routed to implementation level are not mentioned in the NAPCC document. The absence of financial guidelines makes the document toothless. Besides, the lack of mentions of institutional mechanisms of implementing missions made the seriousness of the document futile. As infant document, it left many operational details to concerned departments. The evaluation and monitoring of the mission objectives were also missing at the outset. Although the study finds that at present nearly all states and UTs have submitted their SAPCC to the Union Government having both the adaptation and mitigation strategies underlined in the document yet most of the SAPCCs have been endorsed by the National Steering Committee on Climate Change.

17.4 Summary and Conclusion

The study finds that several missions under the NAPCC are adaptation-oriented which requires adequate state interventions. For instances, among eight missions, the following missions, (a) National Mission on Sustainable Habitat (b) National Water Mission (c) National Mission for Sustaining the Himalayan Ecosystem (d) National Mission for a Green India (e) National Mission for Sustainable Agriculture, and (f) National Mission on Strategic Knowledge for Climate Change, are loaded with adaptation components. The central ministry formulates climate policy and programme implementation mechanism and directs the state to form their actions plan in sync with national objectives and implementation mechanism. Therefore, it can be broadly inferred from the analysis that Centre hardly consults States in environmental/climate-change policymaking, resource allocations, and programme implementation.

To iron out the differences, Centre should include states in the policy perspective by creating a federal council on the model of inter-state council, whose functions can be federally coordinated, and its meeting should regularly be held. The Council will act as the collaborative forum through which states' concerns can be addressed through central policies and programmes. Several Missions of the NAPCC are also reflected in the SAPCCs of states that have to be mainstreamed in various sectoral programmes implemented by the nodal ministries. These programmes need to be revisited as regards to its objectives focusing on addressing climate change issues. Coordination and convergences of programmes and mandates of the institutions should be given top priorities.

Acknowledgements We acknowledge Karnavati University, Gandhinagar and Amity University, Kolkata for providing infrastructure to conduct this research work. Our special thanks to Ministry of Environment, Forest and Climate Change (MoEFCC), for secondary data resources.

References

Agarwal A, Narain S (1991) Global warming in an unequal world: a case of environmental colonialism. Centre for Science and Environment, New Delhi

Dash SK (2007) Climate change: an Indian perspective. Cambridge University Press and Centre for Environmental Education, Ahmadabad

Dubash NK (2016) 13 Climate Change through the Lens of Energy Transformation. New Earth Politics: Essays from the Anthropocene, 315

Government of India (2008) National Action Plan on Climate Change (NAPCC), PM's Council on Climate Change, New Delhi

Held D, Roger C, Nag, EM (2014) Climate governance in the developing world. John Wiley & Sons

INCCA (2010) Climate change and India: A 4x4 assessment—a sectoral and regional analysis for 2030s. Indian Network for Climate Change Assessment, Ministry of Environment and Forests, Government of India, New Delhi, India

IPCC (2007, 2014, 2015) Climate change: synthesis report. Intergovernmental Panel on Climate Change, Geneva, Switzerland

Jogesh A, Dubash NK (2015) State-led experimentation or centrally-motivated replication? A study of state action plans on climate change in India. J Integr Environ Sci 12(4):247–266. https://doi.org/10.1080/1943815X.2015.1077869

Methmann C, Rothe D (2013) Apocalypse now!: From exceptional rhetoric to risk management in global climate politics. In Interpretive Approaches to Global Climate Governance (pp. 105–121). Routledge.

MoEFCC (2015) India: First Biennial Update Report to the United Nations Framework Convention on Climate Change. Ministry of Environment, Forest and Climate Change, Government of India

NATCOM (2004, 2012) Government of India, India's Second National Communication to the United Nations Framework Convention on Climate Change. Ministry of Environment and Forests

PIB, GoI (2013) Executive committee on climate change constituted. GoI, January 31. http://pib.nic.in/newsite/erelease.aspx?relid=91924

Stevenson H, Dryzek JS (2014) Democratizing global climate governance. Cambridge University Press, Cambridge

TERI (2009) Integrating environment, ecology and climate change concerns in the Indian fiscal federalism framework (Project Report No-2007EM04). The Energy and Research Institute, New Delhi

UNEP. Division of Early Warning (2011) UNEP yearbook 2011: Emerging issues in our global environment. UNEP/Earth print

Chapter 18
Carbon Dioxide Reduction in Environment by Biomass Gasification Approach

Gopal Sonkar

Abstract The paper has explored the role of carbon dioxide capture technology (CCt) in minimizing carbon dioxide (CO_2) produced during the power plant process. The evaluation is based on an in-depth review of the literature, including studies from the biomass gasification power plant research and development and academia. The CCt in biomass gasification power plants is often an important option to stabilize CO_2 emissions. The gasification plant captures CO_2 at suitable measures within acceptable limits. One of the major redeeming factors from the gasification process is that CO_2 released into the atmosphere does not exceed what would have been produced. This type of plant provides power to the grid to indirectly prevent a proportional quantum of pollutants. As biomass plants are thus indirectly environment-friendly, they deserve to be an encouragement. The reduction of emissions has the potential to reduce the greenhouse effect. Among emerging CCt, the Absorption technology has been given considerable attention. The gasification with CCt is designed to capture 84% of CO_2.

Keywords Gasification · Carbon dioxide capture technology · CO_2 emission · Absorption · Climate change curtails potential

18.1 Introduction

The United Nations Development Programmes are generally combined with the goal of a sustainable and environment-friendly development mechanism (UNDP 2015). These goals are based on low or zero-carbon energy rcsources combined with ecological systems. Currently, technologies exist to capture CO_2 emissions produced during energy/power making in thermal and biomass gasification plants

G. Sonkar (✉)
Department of Geography, Delhi School of Economics, University of Delhi, Delhi 110007, India
e-mail: gopal.sonkar@gmail.com

A. Dubey et al. (eds.), *Promoting Sustainability Through Water Management and Climate Change Adaptation*, Advances in Geographical and Environmental Sciences,
https://doi.org/10.1007/978-981-99-5479-7_18

(Sandeep et al. 2014). The most affordable technological options are "biomass gasification," in which biomass residues produce electricity in a gas turbine through the thermodynamic process. The use of gasifying gases in a gas turbine limits contaminants (N_2, COS, H_2S, HCN, NH_3, dust, tar, and alkali). CCt capture separates CO_2 from the power plants' emissions that stop entering the atmosphere. CO_2 capture is considered an option in the portfolio of mitigation actions to stabilize atmospheric greenhouse effects. CO_2 emission reductions could be slowed down if CCt is used. Technology maturity, costs, and the ability of developing countries to implement it are important factors. Carbon separation technologies constitute an innovation in specific environmental performance (Butterman and Castaldi 2009).

18.2 Data and Method

In the evaluation process, the research and academic studies of biomass gasification power plant research and development were consulted. It has been done through discourse analysis and content analysis with the help of quantitative and qualitative assessment. Using gasification processes in place of open fire reduces CO_2 emissions is calculated in many studies. We understand the following equation based on experimental and running power plant assessment studies. To calculate the net reduction of CO_2 emissions expressed by R_N in kg/Ce/ TDM (kilograms of Carbon equivalent per tonne of dry Biomass), the formula is as follows:

$$R_N = R_G - R_T$$

R_G is considered the gross decrease in CO_2 emissions in kg/Ce/ TDM. Moving away from traditional fires and toward a more effective technology that reduces CO_2 emissions with the renewable biomass used, such as crop residues and fuelwood. The total amount of CO_2 emitted during biomass power generation as R_T. Figure 18.1 shows the open down-stream gas generator diagram. CO_2 from the flue gas stream in the char zone is captured at around 84% of CO_2 (STAI 2018; MNRE 2017).

18.3 Results

Biomass gasification with CCt has ensured energy sustainability and CO_2 emission reduction. In this scenario, gross CO_2 reduction (R_G) is calculated based on the CO_2 extracted from the Char zone of gasification. There are specific changes in CO_2 emissions from biomass power generation related to the Char zone's origin. A negative axis for R_T has been used to separate it from CO_2 emissions. A conventional mono-ethanol-amine with a CO_2 removal efficiency of 84% is expected to capture the char zone's flue gas mixture.

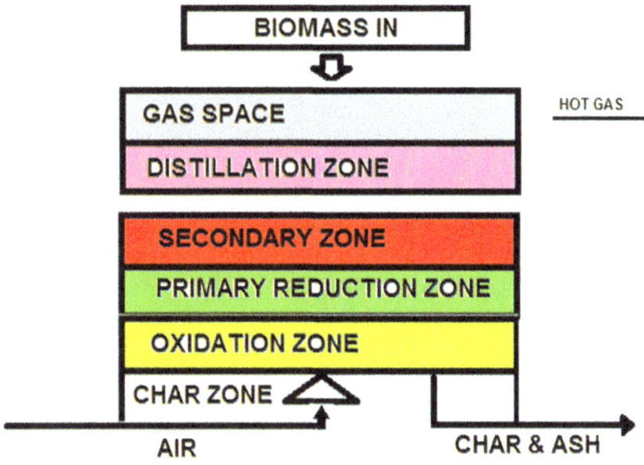

Fig. 18.1 Number of chain chemical reaction inside a gasifier creating the following reaction zones (*Sources* MNRE Reports)

18.4 Discussion

18.4.1 Biomass Gasification Technology in India

Biomass gasification is "converting solid biomass through thermochemical reactions to the production gas." Wood, cotton stems, coconut shells, powdery biomass such as rice husk and sawdust are used in biomass gasification (Dasappa 2017). MNRE (Ministry of New and Renewable Energy) initially pointed out that most of the development of gasification technology is focused on dual fuel systems. The ministry has been introducing biomass gasification through a national biomass gasification program for mechanical electric and thermal heating applications for village electrification since mid-2000. Low-capacity and low-unit-cost engines are among the features that make it easy to adapt to conventional internal combustion engines (MNRE 2017). Gasification involves a two-step response consisting of a process of oxidation and reduction. Substoichiometric air conditions with biomass are required for these processes to take place. Volatiles from biomass were expelled during the first stage of substoichiometric oxidation. Gaseous byproducts, such as carbon dioxide, hydrogen, carbon monoxide, water vapour, and various compounds including N_2, COS, H_2S, HCN, NH_3, dust, tar, and alkali substances, are produced within the temperature range of 1400 to 1500 Kelvin (Yeh and Bai 1999). Several chain chemical reactions are carried out inside a gasifier, creating the following reaction zones.

Updraft gasifiers and downdraft gasifiers are the two primary types of gasifiers, distinguished by the direction in which the producer gas flows within the device (Fig. 18.2).

The biomass descends from the top of the gasifier, while the gases release light rise, resulting in a counter-current movement. Because it contains impurities like tar, up-draft gasifier product gas has an acceptable quality. Because of its impurities, the resultant producer gas has a higher capacity to generate heat when burned, making it ideal for use in heating activities (Dan Langseth 2018). Fuel beds are stratified to keep pollutant levels low by increasing the bed volume. The open down-stream gas

Fig. 18.2 Updraft gasifier (*Sources* MNRE Reports)

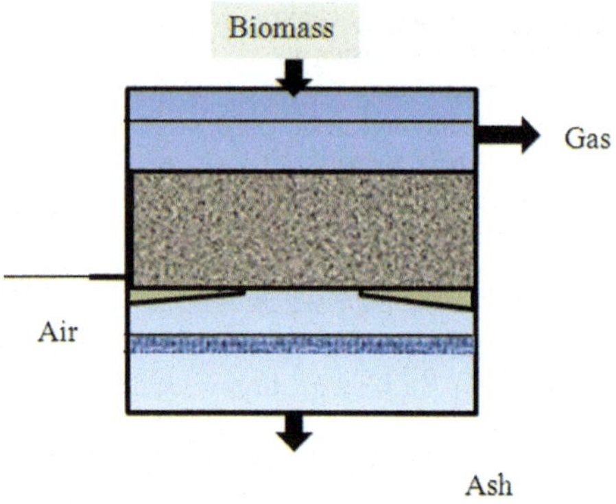

generator diagram is shown in Fig. 18.1. There are six interconnected systems or areas in the gasification process: the gaseous area, the distillation area, the secondary area, the primary production area, the oxidation area, and the char area (see Fig. 18.1). Only the char zone was examined out of the six zones in this research. The carbon-capture facilities can be found in this area. The utilisation of two-bed Skarstrom rings, equipped with pressure equalisation mechanisms derived from the carbon capture unit. Zeolite 13x is an adsorbent that is commonly used. It is estimated that a conventional mono-ethanol-amine (30%) can capture 84% of the CO_2 (Yeh and Bai 1999; Bai and Yeh 1997).

CO_2 capture technology and biomass power infrastructure are expensive and need ample financial support (Mukunda 2018). Some Indian biomass gasification power systems are running without CO_2 capture technology. However, still, these biomass gasification systems mitigate CO_2 emissions. How? That is an important point; see, "in the case of without CCt plants, using agro residues as a fuel in plants does not result in net carbon emissions because the agro residues absorb the CO_2 during the growth phase. (providing production and harvesting is sustainable). Using this energy instead of fossil fuels reduces greenhouse gas emissions and other pollutants, such as sulphur. As coal is the primary energy source for power generation, biomass gasification is assumed to replace coal-based power in environmental impact calculations" (Ravindranath 2013). In India, CO_2 emissions from coal combustion are the primary contributor to climate change. Renewable biomass energy's potential to reduce CO_2 emissions has been widely acknowledged worldwide. According to a study on the potential of biomass power for decentralized applications, nearly 40 million metric tonnes of CO_2 emissions could be avoided by simply switching to biomass gasifiers of 20–200 kW in rural India (Ravindranath and Hall 1995).

During the Biomass Energy for Rural India project, two scenarios were used to evaluate the carbon mitigation potential of the establishment and operation of biomass power plants. The results show that scenario 1 would save 198.44 metric tonnes of carbon dioxide (CO_2) from 2001 to 2010, or 22.04 metric tonnes of CO_2 per year. Under scenario two, CO_2 savings of 374.74 tonnes from 2001 to 2010 equates to an annual savings of around 41.6 tonnes of CO_2 in an optimistic (Ravindranath 2013).

18.4.2 Carbon Dioxide Capture Technology

Carbon Capture Technology has an alternative strategy for addressing CO_2 emissions reduction. There are five technologies available, namely: Absorption (liquid)-MEA, Adsorption CFCMS, Adsorption PSA, Cryogenic distillation, and Membrane Separation. The first is that Absorption (liquid) MEA is economically feasible in Indian gasification practice. The most commonly used carbon capture technology uses chemical solvents, such as mono-ethanol-amine (MEA), as a separation operative. CO_2 separation units installed in biomass gasification plants have been the primary focus of technology studies on upgrading syngas for biofuels (Mukunda 2018).

18.4.3 CO_2 Concentration in Atmosphere and How Importance Role of CCt

The Mauna Loa Observatory in Hawaii has been monitoring CO_2 levels since 1958. In 1958, CO_2 levels were 315 ppm, and in 2010 they were 389 ppm (Sundal et al. 2011). CO_2 levels have risen at a faster rate in recent years (1995–2005, an average annual increase of 1.9 ppm) than they had previously (1.4 ppm from 1960–2005) (Forster 2007). From 2000 to 2009, Loa Mauna's average annual increase was about 2.0 ppm. Scientific research has revealed that burning fossil fuels is a significant factor in increasing CO_2 in the atmosphere. Global CO_2 emissions have experienced a strong acceleration since 2000. A 1.1% annual growth rate was seen in the 1990s related to a 3.3% annual development rate from 2000 to 2006. There was a 2% increase in 2008 over 2007 and a 29% increase between 2000 and 2008 in non-renewable fuel emissions of 8.7 0.5 Gt Cy-1 (Le Quere et al. 2009). The increase is due to a growing global economy and increased fossil fuel consumption per unit of energy used (Canadell 2007b).

CO_2 emissions have shifted from oil to coal as the largest source of CO_2 emissions from this fossil fuel. In 2008, coal accounted for 40% of CO_2 emissions, compared to 37% in the 1990–2000 period. Oil decreased from 41% (1990–2000) to 36% in 2000–2008 (Le Quere et al. 2009). Almost all coal (93%) is burned to produce electricity, while most petroleum (72%) is used to produce transportation fuel (Canadell 2007). Coal generates more CO_2 per unit of energy than petroleum.

The CO_2 deposition in the atmosphere is examined in the preceding section. The impact of carbon dioxide and other greenhouse sequestration creates a global warming situation. The present study selected a CO_2 reduction technology with a biomass gasification power that proved to have climate change reduction potential (CCRP). As a result of introducing CCt into the biomass gasification plant, CO_2 emissions can be reduced. Reduced CO_2 emissions at gasification plants to generate the required operation of CCt because of the environmental footprint associated with carbon. According to an estimate, in the absence of the CCt, about 90% of the CCRP is due to CO_2 emissions from power plants and other industries (Mukunda 2018). The contribution of CCt to the CCRP through the gasification system and the proportion of major greenhouse gases is significant. As a result of the power plant's energy conversion, CO_2 emissions account for 90% of the CCRP (10.9% of impact) (Horne 2019). For gasification plants where the CCt process is embedded, a reduction in CO_2 emissions is reported (BEI 2018).

18.5 Conclusions

The integration of CCt into the biomass gasification plant would reduce CO_2 through substantial net sequestration of CO_2 in the atmosphere. The existing gasification biomass power plants are small- and medium-scale heat, and power generation with the installation of CO_2 capture technology is significant. This work analyzed the environmental performance of CCt to be applied in biomass gasification plants. An efficient CO_2 sequestration configuration is considered the best practice.

References

Bai H, Yeh AC (1997) Removal of CO_2 greenhouse gas by ammonia scrubbing. Ind Eng Chem Res 36(6):2490–2493

BEI (2018) Modern power station practice. British Electricity International, London

Butterman HC, Castaldi MJ (2009) CO_2 as a carbon neutral fuel source via enhanced biomass gasification. Environ Sci Technol 43(23):9030–9037

Canadell JG (2007a) Contributions to accelerating atmospheric CO_2 growth from economic activity, carbon intensity, and efficiency of natural sinks. Natl Acad Sci 104(47):18866–18870

Canadell JG (2007b) Factoring out natural and indirect human effects on terrestrial carbon sources and sinks. Environ Sci Policy 10(4):370–384

Dan Langseth GL (2018) Gasification technology. Black, London

Dasappa, S. (2017). Experimental and modeling: Indian Institute of Technology, Mumbay. MNRE, New Delhi

Forster PV (2007) Changes in atmospheric constituents and in radiative forcing. In: Climate change 2007: the physical science basis, pp 131–234

Horne B (2019) Power plants bio-fuels made simple. MNRE, New Delhi

Le Quere C et al (2009) Trends in the sources and sinks of carbon dioxide. *Nature Geoscience* 2:831–839. https://doi.org/10.1038/ngeo689. Published online 17 Nov. 2009

MNRE (2017) Industrial energy and electric power from biomass residues by Indian Institute of Technology, Mumbay. MNRE, New Delhi

Mukunda PP (2018) Recent advances in biomass gasification and combustion. MNRE, New Delhi

Ravindranath D (2013) Biomass energy for rural India carbon mitigation report. Biomass Energy For Rural India Society, Bangalore

Ravindranath NH, Hall DO (1995) Biomass, energy and environment. Oxford University Press, London

Sandeep K, Snehesh S, Dasappa S (2014). Carbon dioxide capture through biomass gasification. Indian Institute of Science, Bangalore, 1 to 7

STAI (2018) Combined factory/energy workshop efficient production and utilization of steam in sugar factories. In: The sugar technologists association of India (p. Papers for the presentation at the ISSCT Combined Factory/Energy Workshop Efficient Production and Utilization of Steam in Sugar Factories). The Sugar Technologists Association of India, Pune
Sundal AV, Shepherd A, Nienow P, Hanna E, Palmer S, Huybrechts P (2011) Melt-induced speed-up of Greenland ice sheet offset by efficient subglacial drainage. Nature 469(7331):521–524
UNDP (2015) United nations development programme and climate change: zero carbon, sustainable development. United Nations Development Programme, New York
Yeh AC, Bai H (1999) Comparison of ammonia and monoethanolamine solvents to reduce CO_2. Sci Total Environ 228(2–3):121–133

Correction to: A Change Detection Analysis of Mangrove Forests in and Around Devi River Mouth, Odisha Using Remote Sensing and GIS Technique

Prasanna Kumar Nayak and Sujata Mishra

Correction to:
Chapter 4 in: A. Dubey et al. (eds.), *Promoting Sustainability Through Water Management and Climate Change Adaptation*, Advances in Geographical and Environmental Sciences, https://doi.org/10.1007/978-981-99-5479-7_4

The original version of the book was inadvertently published without the references in Chapter 4, which have now been included. The correction chapter has been updated with the changes.

The updated version of this chapter can be found at
https://doi.org/10.1007/978-981-99-5479-7_4

A. Dubey et al. (eds.), *Promoting Sustainability Through Water Management and Climate Change Adaptation*, Advances in Geographical and Environmental Sciences,
https://doi.org/10.1007/978-981-99-5479-7_19

The manufacturer's authorised representative in the EU is Springer Nature Customer Service Centre GmbH, Europaplatz 3, 69115 Heidelberg, Germany. If you have any concerns regarding our products, please contact ProductSafety@springernature.com

Printed and bound by CPI Group (UK) Ltd, Croydon, CR0 4YY
01/10/2026
02225867-0003